NO EASY ANSWERS

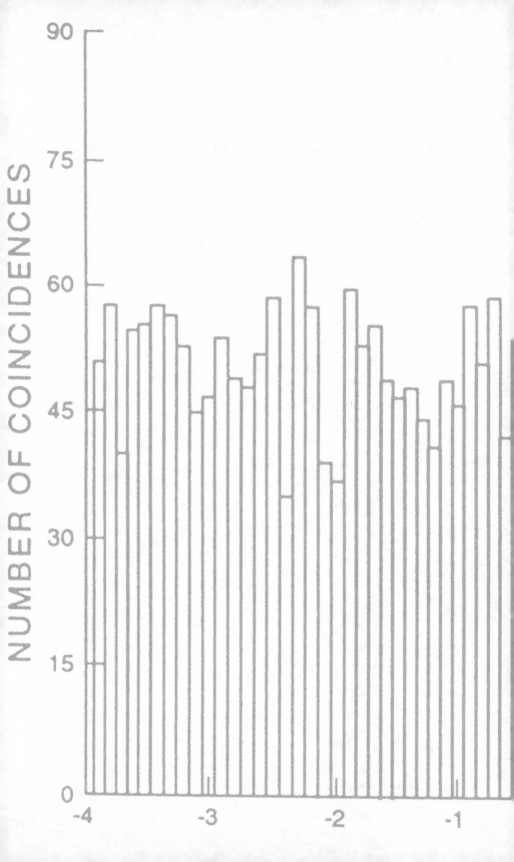

NO EASY ANSWERS

Science and the Pursuit of Knowledge

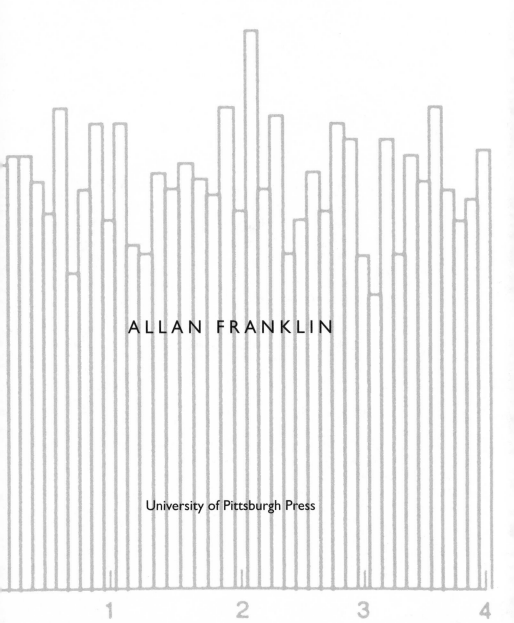

ALLAN FRANKLIN

University of Pittsburgh Press

Q162
F694

Published by the University of Pittsburgh Press, Pittsburgh, Pa. 15260
Copyright © 2005, University of Pittsburgh Press
All rights reserved
Manufactured in the United States of America
Printed on acid-free paper
10 9 8 7 6 5 4 3 2 1

Library of Congress Cataloging-in-Publication Data
Franklin, Allan, 1938-
No easy answers : science and the pursuit of knowledge / Allan Franklin.
  p. cm.
Includes bibliographical references and index.
ISBN 0-8229-4250-X (acid-free paper)
1. Science—Popular works. 2. Science—Methodology—Popular works. I. Title.
Q162.F694 2005
507'.2—dc22
                                        2004015715

Contents

List of Illustrations

Figures

Tables

Preface

This book is intended to give an accurate picture of science to both a general reader and my colleagues in the humanities and social sciences, who may not have any background in physics. I present several case studies from the history of recent physics and one from molecular biology that illustrate the various roles that experiment plays in science. These cases show that experiment plays legitimate and important roles in science and that it provides the basis of our knowledge of the physical world. In short, I argue that science is a reasonable enterprise based on valid experimental evidence and on reasoned and critical discussion.

Most of the examples are from physics, because that is the science I know best, but I believe that the episodes presented are typical of all science. I also show that the strategies that I suggest are used to validate an experimental result are used in both physics and evolutionary biology. I must make one disclaimer. Presenting an accurate picture of the practice of physics requires the inclusion of some technical content. This includes both graphs of the experimental results and, very rarely, an equation. The reader can be assured, however, that there will be no examination. These studies should be read as stories—evidential fables, perhaps. This will allow the reader to understand the important elements of the episodes. However, the term fables should not imply that these case studies are in any way fanciful or imagined. They are accurate and somewhat simplified accounts of the actual history, which are intended to illustrate the roles that experiment legitimately plays in science.

Because these are necessarily condensed histories, I include footnotes on the first page of most chapters that refer readers to a more detailed study. I also provide a glossary of technical terms.

The case studies in this book were compiled over a period of twenty-five years. During that time I have been helped by so many friends and colleagues that merely listing all of them and the assistance they have provided would add another chapter to this book. I believe that all of them have been acknowledged in my previous work, so I will only say that this work could not have been done without their help and encouragement. I also thank Valerie Melendez, Jessica Rodriguez, Tom Lyons, and Jeffrey Robinson for helpful comments on parts of the manuscript. Last, and certainly not least, I thank my wife, Cynthia Betts, without whom none of this would have been possible.

NO EASY ANSWERS

Introduction

In 2000, *Time* magazine selected Albert Einstein as "Person of the Century." He was chosen over political figures, athletes, and entertainers. This was not only a recognition of Einstein's contributions to science, but also evidence of the high regard for science in the United States. As George Levine wrote about his own book (1986, 24), "It takes seriously the view that science is one of the great achievements of the human mind, that it matters powerfully to us, for better or worse, in the way we live, the way we think, and the way we imagine. There is no literature more important."

Not everyone agrees. Harry Collins, whose work is discussed in chapter 13, has compared science to a golem, a mythical creature made of clay and water by means of spells and incantations (Collins and Pinch 1993). The golem is clumsy at best, evil at worst. Collins further remarks that "the natural world has a small or non-existent role in the construction of scientific knowledge" (1981). Andrew Pickering commented that "there is no obligation upon anyone framing a view of the world to take account of what twentieth-century science has to say" (1984b, 413). These views, known as social constructivism, form part of what has been called the postmodern critique of science. Whether it is because of sexism, Eurocentrism, or the interests of scientists, science is severely, if not fatally, flawed according to these appraisals.

I disagree. I believe that science is a reasonable enterprise that provides us with knowledge of the natural world that is based on valid experimental evidence and reasoned and critical discussion. I argue that nature, as revealed by experiment, plays an important and legitimate role in science. As the late Richard Feynman, one of the leading theoretical physicists of the 20th century, wrote, "The principle of science, the definition, almost, is the following: *The test of all knowledge is experiment. Experiment is the sole judge* of scientific truth" (Feynman, Leighton, and Sands 1963, I–1). Today this might seem to be an old-fashioned view, but I believe it is well supported by the evidence.

To persuade my readers that this view is correct, I cannot, however much I might like to, provide a tour of a laboratory to show an experiment performed from beginning to end and reveal how results are produced, or arrange access to scientists' discussions concerning the acceptance or rejection of theories. Instead, case studies of episodes from recent science illustrate the various roles of experiment in science. These are primarily from the field of physics, the science I know best, but I believe they are typical of all science.

Experiment plays many roles in science. One of its important roles is to test theories and to provide the basis for scientific knowledge. It can also call for a new theory, either by showing that an accepted theory is incorrect or by exhibiting a new phenomenon that needs explanation. Experiment can provide hints toward the structure or mathematical form of a theory, and it can provide evidence for the existence of the entities involved in theories. It can also measure quantities that theory maintains are important. Finally, it may also have a life of its own, independent of theory. Scientists may investigate a phenomenon just because it looks interesting. This also provides evidence for a future theory to explain.

It must be remembered, however, that science is fallible. Theoretical calculations, experimental results, and the comparison between experiment and theory may all be wrong. Science is more complex than "the scientist proposes, Nature disposes." It may not always be clear what the scientist is proposing. Theories often need to be articulated and clarified. It also may not be clear how Nature is disposing. Experiments may not always give clear-cut results and may even disagree for a time. Sometimes they can be incorrect.

An Epistemology of Experiment

If experiment is to play all these important roles in science and to provide the evidential basis for scientific knowledge, then good reasons to believe in the results are required. It has been more than two decades since Ian Hacking asked, "Do we see through a microscope?" (1981). Hacking is really asking, How do we come to believe in an experimental result obtained with a complex experimental apparatus? How do we distinguish between a valid result and an artifact created by that apparatus? Hacking also provides an extended answer (1983). He points out that even though an experimental apparatus is laden with, at the very least, the theory of the apparatus, observations remain robust despite changes in the theory of the apparatus or in the theory of the phenomenon. His illustration is the continuous belief in microscope images despite the major change in the theory of the microscope when Abbe pointed out the importance of diffraction in its operation. One reason Hacking gives for this is that in making such observations the experimenters intervened. They manipulated the object under observation. Thus, the view of a cell through a microscope should be altered by injecting fluid into the cell or staining the specimen. The cell changes shape or color when this is done. Observing the predicted effect strengthens belief in both the proper operation of the microscope and the observation. This is true in general. Observing the predicted effect of an intervention strengthens belief in both the proper operation of the experimental apparatus and the observations made with it.

Hacking also discusses the strengthening of belief in an observation by independent confirmation. The fact that the same pattern of dots, dense bodies in cells, is seen with different microscopes—that is, ordinary, polarizing, phase-contrast, fluorescence, interference, electron, acoustic, and so on—argues for the validity of the observation. Hacking argues that it would be a preposterous coincidence if the same pattern of dots were produced in two totally different kinds of physical systems. Different apparatuses have different backgrounds and systematic errors, making the coincidence, if it is an artifact, most unlikely. If it is a correct result, and the instruments are working properly, the agreement of results is understandable.

Hacking's answer is correct as far as it goes, but it is incomplete. On occasion an experiment can be performed with only one type of apparatus, such as an electron microscope or a radio telescope, or intervention is either impossible or extremely difficult. Other strategies are then needed to validate the observation. These may include

1. **Using experimental checks and calibration, in which the experimental apparatus reproduces known phenomena.** For example, to ascertain whether the spectrum of a substance obtained with a new type of spectrometer is correct, an experimenter might check whether the new spectrometer could reproduce the known spectrum of visible light emitted by hydrogen, the Balmer series. If the Balmer series is observed (as expected), then belief that the spectrometer is working properly is affirmed. This also strengthens belief in the results obtained with that spectrometer. If the check fails, then there is good reason to question the results obtained with that apparatus.

2. **Reproducing artifacts that are known in advance to be present.** An example of this comes from experiments to measure the infrared spectra of organic molecules. It was not always possible to prepare a pure sample of such material. Sometimes the substance had to be placed in an oil paste or in solution. In such cases, two superimposed spectra were observed: the spectrum of the substance and the spectrum of the oil or the solvent The latter was then compared with the known spectrum of the oil or the solvent. Observation of this artifact gave confidence in other measurements made with the spectrometer.

A somewhat different and interesting example of this occurred in the 1980s in a trial in Boulder, Colorado. The defendant was convicted of heroin possession on the basis of infrared spectroscopy of a white powder found in his possession. The spectrum of the powder matched that of heroin. An expert witness for the defense raised the legitimate objection that the spectrometer had not previously been run without a sample present to verify that there had been no prior contamination of the instrument. On reading this in a local newspaper, I reflected that if the instrument had been contaminated with heroin, and the trial substance was something other than heroin, then what would have been observed was the spectrum of the other substance superimposed on the heroin spectrum. After the trial was over, I asked a member of the jury if my argument had been considered. He told me that although the jury had considered the defense objection seriously, they opted for conviction because the defense had had ample opportunity to have the substance independently tested and had not done so. They inferred that the defense knew the substance was heroin and had avoided the test.

3. **Eliminating plausible sources of error and alternative explanations of the result (the Sherlock Holmes strategy).** As Holmes remarked to Watson in *The Sign of Four*, "How often have I said to you that when you have eliminated the impossible, whatever remains, *however improbable*, must be the truth." Thus, when scientists claimed to have observed electric discharges in the rings of Saturn, they argued for their result by showing that it could not have been caused by defects in the telemetry, by interaction with the environment of Saturn, by lightning, or by dust. The only remaining explanation of their result was that it was due to electric discharges in the rings. There was no other plausible explanation of the observation. In addition, the same result was observed by both *Voyager 1* and *Voyager 2*. This provided independent confirmation. Often, several epistemological strategies are used in the same experiment.

4. **Using the results themselves to argue for their validity.** Galileo's telescopic observations of the moons of Jupiter are an example. Although it was quite plausible that his early telescope might have created spots of light, it was extremely implausible that the telescope would create them so that they would appear to be a small planetary system with eclipses and other consistent motions. It was even more implausible that the created spots would satisfy Kepler's third law: $R^3/T^2 = $ constant, where R is the radius of a satellite's orbit and T is its period. (Kepler's third law was not available when Galileo made his observations, but it helps to validate them.) A similar argument was used by Robert Millikan to support his

observation that electric charge came only in integral multiples of a fundamental unit of charge and his measurement of the charge of the electron. Millikan remarked (1911), "The total number of changes which we have observed would be between one and two thousand, and *in not one single instance has there been any change which did not represent the advent upon the drop of one definite invariable quantity of electricity or a very small multiple of that quantity.*" In both of these cases the argument is that there was no plausible malfunction of the apparatus, or background, that would explain the observations.

5. **Using an independently well-corroborated theory of the phenomena to explain the results.** This was illustrated in the discovery of the $W^\pm$, the charged intermediate vector bosons required by the Weinberg-Salam unified theory of electroweak interactions. Although these experiments used very complex apparatuses and used other epistemological strategies, the agreement of the observations with the theoretical predictions of the particle properties helped to validate the experimental results. In this case the particle candidates were observed in precisely the type of events predicted by the theory. In addition, the measured particle mass of 81 ± 5 GeV (billion electron volts)/c^2 and 80^{+10}_{-6}, GeV/c^2, found in two experiments (the independent confirmation should be noted), was in good agreement with the theoretical prediction of 82 ± 2.4 GeV/c^2. It was improbable that any background effect, which might mimic the presence of the particle, would be in such good agreement with theory.

6. **Using an apparatus whose proper operation is predicted by a well-corroborated theory.** In this case the support for that theory passes on to the apparatus. This is the case with both the electron microscope and the radio telescope, whose proper operation is predicted by a well-supported theory, although other strategies are also used to validate the observations.

7. **Using statistical arguments.** An interesting example of this arose in the 1960s when the search for new particles and resonances occupied a substantial fraction of the time and effort of those physicists working in experimental high-energy physics. The usual technique was to plot the number of events observed as a function of the invariant mass of the final-state particles and to look for bumps above a smooth background. The usual informal criterion for the

presence of a new particle was that it resulted in a three-standard-deviation effect above the background, a result that had a probability of 0.27% of occurring in a single bin. This criterion was changed to four standard deviations, which had a probability of 0.0064%, when it was pointed out that the large number of graphs plotted each year by high-energy physicists made it rather probable, on statistical grounds, that a three-standard-deviation effect would be observed.

These strategies, along with Hacking's intervention and independent confirmation, constitute an epistemology of experiment. They provide good reasons for belief in experimental results. They do not, however, guarantee that the results are correct. In some experiments these strategies are applied, but the results are later shown to be incorrect. Experiment is fallible. Neither are these strategies exclusive or exhaustive. No single one of them, or set of them, guarantees the validity of an experimental result. Physicists use as many of the strategies as they can in any given experiment.

Although all these illustrations of the epistemology of experiment come from physics, David Rudge (1998, 2001) has shown that they are also used in biology. His example is Bernard Kettlewell's evolutionary biology experiments on the peppered moth, *Biston betularia* (1955, 1956, 1958). The typical form of the moth has a pale, speckled appearance, and there are two darker forms, *f. carbonaria*, which is nearly black, and *f. insularia*, which is intermediate in color. The typical form of the moth was most prevalent in the British Isles and Europe until the middle of the 19th century. At that time things began to change. Increasing industrial pollution had both darkened the surfaces of trees and rocks and killed the lichen cover of the forests downwind of pollution sources. Coincident with these changes, naturalists had found that rare, darker forms of several moth species, in particular the peppered moth, had become common in areas downwind of pollution sources.

Kettlewell attempted to test a selectionist explanation of this phenomenon. E. B. Ford (1937, 1940) had suggested a two-part explanation of this effect: (1) darker moths had a superior physiology, and (2) the spread of the melanic gene was confined to industrial areas because the darker color made *carbonaria* more conspicuous to avian predators in rural areas and less conspicuous in polluted areas. Kettlewell believed that Ford had established the superior viability of darker moths, and he wanted to test the hypothesis that the darker form of the moth was less conspicuous to predators in industrial areas.

Kettlewell's investigations consisted of three parts. In the first, he used human observers to investigate whether his proposed scoring method would be accurate in assessing the relative conspicuousness of different types of moths against different backgrounds. The tests showed that moths on "correct" backgrounds—typical on lichen-covered backgrounds and dark moths on soot-blackened backgrounds—were almost always judged inconspicuous, whereas moths on "incorrect" backgrounds were judged conspicuous.

Next, Kettlewell released birds into a cage containing all three types of moth and both soot-blackened and lichen-covered pieces of bark as resting places. After some difficulties, he found that birds prey on moths in an order of conspicuousness similar to that gauged by human observers.

Finally, Kettlewell investigated whether birds preferentially prey on conspicuous moths in the wild. He conducted a mark-release-recapture experiment in a polluted environment (Birmingham) and later in an unpolluted wood. He released 630 marked male moths of all three types in an area near Birmingham, which contained predators and natural boundaries. He then recaptured the moths with two different types of trap, each containing virgin females of all three types to guard against the possibility of pheromone differences.

Kettlewell found that *carbonaria* was twice as likely to survive in soot-darkened environments (27.5 percent) as the typical moth (12.7 percent). He worried, however, that his results might be an artifact of his experimental procedures. Perhaps the traps used were more attractive to one type of moth, one form of moth was more likely to migrate, or one type of moth just lived longer. He eliminated the first alternative by showing that the recapture rates were the same for both types of trap. The use of natural boundaries and traps placed beyond those boundaries eliminated the second possibility, and previous experiments had shown no differences in longevity. Further experiments in polluted environments confirmed that *carbonaria* was twice as likely to sur-

TABLE I
Examples of epistemological strategies used by experimentalists in evolutionary biology

Epistemological strategies	Examples from Kettlewell
1. Experimental checks and calibration, in which the apparatus reproduces known phenomena	Use of the scoring experiment to verify that the proposed scoring methods would be feasible and objective
2. Reproduction of artifacts that are known in advance to be present	Analysis of recapture figures for endemic *betularia* populations
3. Intervention, in which the experimenter manipulates the object under observation	Not present
4. Independent confirmation using different experiments	Use of two different types of traps to recapture the moths
5. Elimination of plausible sources of background and alternative explanations of the result	Use of natural barriers to minimize migration
6. Use of the results themselves to argue for their validity	Filming the birds preying on the moths
7. Use of an independently well-corroborated theory of the phenomenon to explain the results	Use of Ford's theory of the spread of industrial melanism
8. Use of an apparatus based on a well-corroborated theory	Use of Fisher, Ford, and Shepard techniques [The mark-release-capture method had been used in several earlier experiments]
9. Statistical arguments	Use and analysis of large numbers of moths

Examples are from Bernard Kettlewell's investigations of industrial melanism (1955, 1956, 1958). Reproduced from Rudge (1998).

vive as the typical moth. An experiment in an unpolluted environment showed that the typical moth was three times as likely to survive as *carbonaria*. Kettlewell concluded that such selection was the cause of the prevalence of *carbonaria* in polluted environments.

Rudge also demonstrates that the strategies used by Kettlewell are those I describe. His examples are given in table 1.

The Case Studies

The case studies presented in this book do not relate an unbroken string of successes. Real science is quite different from the idealized science of textbooks. Some cases—such as the discovery of parity non-conservation (the violation of mirror symmetry) in weak interactions and the discovery of the electron, both considerable achievements—are still regarded as correct. Others show the difficulty of arriving at a correct answer and the length of time the process may take. Both the road to the neutrino—how physicists came to realize the need for a new elementary particle—and the solution of the solar neutrino problem took some 30 years. There are also unfinished stories, such as the search for magnetic monopoles.

There are also examples of incorrect theories or hypotheses, such as the Konopinski-Uhlenbeck theory of β-decay. Incorrect experimental evidence suggested the need for an alternative to Enrico Fermi's theory of β-decay. That alternative theory was provided by Emil Konopinski and George Uhlenbeck. Further experimental evidence and an incorrect theory-experiment comparison seemed to support the K-U theory. When these problems were found and corrected, Fermi's theory was better supported.

In the case of the fifth force, a proposed modification of Newton's law of gravity, the first two experimental tests of the hypothesis gave conflicting answers. One experiment supported the fifth force, and the other found no evidence for it. Eventually the discord was resolved, and the physics community came to reasonably believe that there was no such force.

Other examples of discordant results and their resolution include the early searches for gravity waves and the experiments on atomic-parity violation and their relation to the Weinberg-Salam unified theory of electroweak interactions. Both of these cases are presented with two perspectives, one offered by a critic of science.

There are even cases of scientists behaving badly. In the episode of Millikan's measurement of the charge of the electron, the researcher was selective both in the data he used and in his calculational procedures. The effects of his cosmetic surgery were quite small, however, and a safeguard was provided by the numerous subsequent measurements of this important physical quantity. Scientific deception is also represented, by Emil Rupp's fraudulent experiments on electron scattering. Here too, numerous repetitions showed that his results were incorrect.

Despite these errors, wrong turns, dead ends, and misbehavior, these case studies show that experimental evidence plays—as it should—a legitimate role in the production of scientific knowledge. It may take some time for problems to be solved or errors to be corrected, but the histories show that they will be solved and corrected. Science, sooner or later, accurately describes the natural world and its wonders.

Parent

I

EXPERIMENT—MAKING OR

BREAKING THEORIES

F_1

F_2

(2)

The Violation of Parity Conservation

One of the important purposes of experiment is testing theories or hypotheses. One example of a crucial experiment, which decided unequivocally between two competing theories, was the discovery that parity conservation—also known as mirror-reflection symmetry or left-right symmetry—is violated in weak interactions. It is perhaps the clearest case of a crucial experiment in the history of physics. This case is fascinating because experiments done in the late 1920s and early 1930s in retrospect also demonstrate parity nonconservation. The significance of these experiments was not realized by either the experimenters themselves or anyone else in the physics community. It was only after parity nonconservation had been discovered in the 1950s that physicists recognized the significance of the earlier experiments.

Discovery of Parity Nonconservation

In 1956 Tsung Dao Lee and Chen Ning Yang, who would win the Nobel Prize for their suggestion, proposed that parity, or mirror-reflection

For details, see Franklin (1986, chap. 1 and 2).

symmetry or left-right symmetry, is not conserved in weak interactions. (Physicists identify four interactions. In decreasing order of strength they are the strong, or nuclear, interaction, which holds the atomic nucleus together; the electromagnetic interaction, which holds atoms together; the weak interaction, responsible for radioactive decay; and the gravitational interaction.)

Parity conservation was a well-established and strongly believed principle of physics. As students of introductory physics learn, to determine the magnetic force between two currents, first determine the direction of the magnetic field caused by the first current with a right-hand rule and then determine the force exerted on the second current by that field with a second right-hand rule. Exactly the same answer is reached if two left-hand rules are used. This is left-right symmetry, or parity conservation, in electromagnetism and in classical physics.

In 1927 Eugene Wigner proposed the concept of parity conservation in quantum mechanics as a way of explaining some recent results in atomic spectra. It quickly became an established principle. As Hans Frauenfelder and Ernest Henley stated (1975, 359), "Since invariance under space reflection is so appealing (why should a left- and right-handed system be different?), conservation of parity quickly became a sacred cow." An early indication of this came in 1933 when Wolfgang Pauli rejected a theory proposed by Herman Weyl on the grounds that it was not invariant under reflection, or because it did not conserve parity. "However, as the derivation shows, these wave equations are not invariant under reflections (interchanging left and right) and *thus are not applicable to physical reality*" (1933). Pauli, a future Nobel Prize winner for his exclusion principle, a crucial element in the explanation of atomic structure, was notoriously critical and skeptical. He is said to have commented on a paper by another physicist that "it is not even wrong." In this instance, Pauli himself was mistaken.

In the early 1950s, physicists were faced with a problem known as the τ-θ puzzle. According to one set of criteria, that of mass and lifetime, two elementary particles (the τ and the θ) appeared to be the same, whereas by another set of criteria, that of spin and intrinsic parity, they appeared to be different, a very unusual situation in physics. (A good analogy to spin is the rotation of the earth on its axis. Intrinsic parity refers to the mirror-symmetry properties of the wave function, a mathematical function used to describe the particle.) In 1956 Lee and Yang realized that the problem would be solved, and that the two particles would just be different decay modes of the same particle, if parity were not conserved in the decay of the τ- and θ-particles, a weak interaction (Franklin 1986, chap. 1 and 2). They examined the evidence for parity conservation and found, to their surprise, that although there was strong evidence that parity was conserved in the strong interaction and in the electromagnetic interaction, there was, in fact, no supporting evidence that it was conserved in the weak interaction. It had never been tested. It had just been assumed.

The survey by Lee and Yang was incomplete. They overlooked experiments done in the 1920s and 1930s that, in retrospect, provide evidence for parity nonconservation, although no one at the time realized their implications. They also did not find an amusing early test of parity conservation. In their paper, "Movement of the Lower Jaw of Cattle During Mastication," Pascual Jordan and Ralph Kronig (1927) noted that the chewing motion of cows is not straight up and down, but is either a left-circular or right-circular motion. (These motions reverse in a mirror and are also visible in humans.) The results of their survey of cows in Sjaelland, Denmark, indicated that 55% were right circular and 45% were left circular, a ratio they regarded as consistent with parity conservation.

> This nomenclature is based on the tacit assumption that one and the same cow always maintains its sense of rotation. We could confirm this by a limited number of observations but are aware that more complete data, extending over longer periods of time, are necessary to definitely to settle this point. Statistical investigations on cows distributed over the northern part of Sjaelland, Denmark, led to the result that about fifty-five percent were right-circular, the rest left-circular animals. As one sees, the ratio of the two is approximately unity. The number of observations was, however, scarcely sufficient to make sure if the deviations from unity is real. Naturally these determinations allow no generalisation with regard to cows of different nationality.

The physics community generally, and many leading physicists, did not believe that the Lee and Yang suggestion was correct. Pauli skeptically remarked, "I do not believe the Lord is a weak left-hander, and I am ready to bet a very large sum that the experiments will give symmetric results" (quoted in Bernstein, 1967, 59). There were other bets between physicists. Richard Feynman, one of the leading theoretical physicists of the 20th century and a Nobel Prize winner, bet Norman Ramsey, another winner, $50–$1 that parity would be conserved. Ramsey notes that Feynman believed that the real odds were 1 million to 1, but wouldn't bet that much on anything (personal communication). Felix Bloch, yet another Nobel Prize winner, offered to bet his hat with any other member of the Stanford physics department that parity would be conserved (personal communication from T. D. Lee, 1977).

Lee and Yang (1956) suggested several possible experimental tests of parity conservation in the weak interaction. Of the two most important ones, the first was the β-decay of oriented nuclei. (β-decay is the transformation of one atomic nucleus into a different nucleus, with the emission of an electron and a neutrino. Oriented nuclei are nuclei whose spins all point in the same direction.) The second was the sequential decay $\pi \to \mu \to e$. This is the decay of a π meson, an elementary particle, into a μ meson, another particle, and a neutrino. The μ meson subsequently decays into an electron and two neutrinos. These were the first experiments done and provided the crucial evidence for the physics community.

Figure 2.1 helps to explain this. An example is a radioactive nucleus, whose spin points upward and which always emits an electron in the direction opposite to the spin. In the mirror the spin is reversed, whereas the electron's direction of motion is unchanged. Now the electron is emitted in the same direction as the spin. The mirror result is different from the real result. This violates mirror symmetry and shows the nonconservation of parity. Parity conservation would also be violated if, in a collection of oriented nuclei, more electrons were emitted in the direction of the nuclear spin than opposite to the spin, or vice versa. Only if a collection of nuclei emitted equal numbers symmetrically with respect to the spin direction would parity be conserved. For $\pi \to \mu \to e$ decay, parity nonconservation implies that the muon (the μ) will be longitudinally polarized, which means that its spin will point either parallel to or antiparallel to its direction of motion. If the muon is

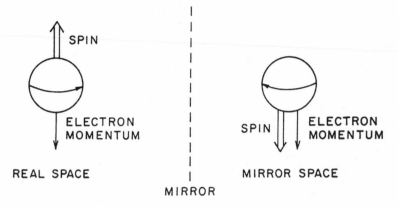

SPIN

ELECTRON
MOMENTUM

REAL SPACE

MIRROR

SPIN

ELECTRON
MOMENTUM

MIRROR SPACE

FIGURE 2.1. Spin and momentum in real space and mirror space. In real space the spin and momentum point in opposite directions. In mirror space they point in the same direction. This is an example of parity nonconservation.

stopped, its polarization remains and its subsequent decay will look just like that of an oriented nucleus. In this case an asymmetry would be expected in the distribution of the muon decay electrons emitted along the direction of the muon motion and opposite to that direction.

The first experiment was performed by Chien-Shiung Wu and her collaborators. It consisted of a layer of oriented ^{60}Co nuclei and a single, fixed, electron counter, which was located either along the direction of, or opposite to, the orientation of the nuclei. The direction of the orientation of the nuclei could be changed and any difference in counting rate in the fixed electron counter observed. The results are shown in figure 2.2. With the counter opposite to the nuclear orientation, the ratio of the counts observed when the nuclei were oriented to when they were not was 1.20. With the counter parallel to the orientation, the ratio was 0.80. If parity were conserved, the ratio would have been one. (In statistical terms this was a 13-standard-deviation effect. This meant that it was extremely unlikely that the observed effect was due to a statistical fluctuation in the number of counts.) This was a clear asymmetry. The experimenters concluded, "If an asymmetry between θ and $180° - \theta$ (where θ is the angle between the orientation of the parent nuclei and the momentum of the electrons) is observed, it provides unequivocal proof that parity is not conserved in β-decay. This

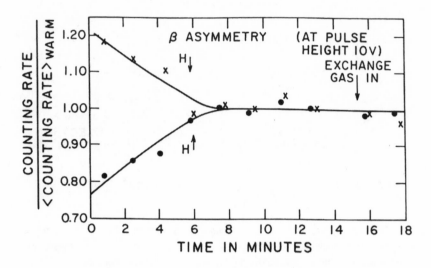

FIGURE 2.2. Relative counting rates for β-particles from the decay of oriented
^{60}Co nuclei for different nuclear orientations (field directions). More electrons
are emitted opposite to the nuclear orientation than in the same direction. This
demonstrates parity nonconservation. From Wu et al. (1957).

asymmetry has been observed in the case of oriented ^{60}Co" (Wu et
al. 1957).

The second experiment, on the sequential decay $\pi \rightarrow \mu \rightarrow e$, was performed with two different experimental techniques by Richard Garwin,
Leon Lederman, and Marcel Weinrich (1957) and by Jerome Friedman
and Valentine Telegdi (1957). The Garwin experiment found a sinu-
soidal variation in counting rate, in contrast to the symmetric distribution expected if parity were conserved. Their statistically overwhelming
effect (22 standard deviations) led them to conclude that parity was
not conserved. In fact, Lederman called Lee at 7 a.m. and announced,
"Parity is dead" (quoted in Lee 1971). Friedman and Telegdi performed
the same experiment with a different technique. They found a forward-
backward asymmetry of 0.091 ± 0.021 (a four-standard-deviation
effect). Like Garwin and colleagues, Friedman and Telegdi also concluded that parity was not conserved.

The immediate reaction of the physics community was that parity
nonconservation in the weak interaction had been clearly demonstrated. It is fair to say that any physicist, upon seeing these experi-

mental results, believed that parity wasn't conserved. Even Pauli was convinced. He wrote (quoted in Bernstein 1967, 60), "Now, after the first shock is over, I begin to collect myself. Yes, it was very dramatic. On Monday, the twenty-first, at 8 p.m. I was to give a lecture on the neutrino theory. At 5 p.m. I received the three experimental papers. I am shocked not so much by the fact that the Lord prefers the left hand, as by the fact that He still appears to be left-right symmetric when He expresses Himself strongly. In short, the actual problem now seems to be the question: Why are strong interactions right and left symmetric?"

Pauli was fortunate that he had not wagered a very large sum of money that parity would be conserved. Feynman paid Ramsey. Bloch remarked that it was lucky he didn't own a hat (personal communication from T. D. Lee, 1977). The Nobel Prize in physics was awarded to Lee and Yang in 1957, less than a year after their suggestion that parity wasn't conserved.

This is, perhaps, the clearest example of a crucial experiment, one that decides unequivocally between two theories or, in this case, between two classes of theory, in the history of physics. The evidence was beyond a reasonable doubt. Three different experiments were conducted, involving two different processes, the β-decay of oriented nuclei and $\pi \rightarrow \mu \rightarrow$ e decay. The statistical evidence was overwhelming. Friedman and Telegdi found a 4-standard-deviation effect, Wu and collaborators a 13-standard-deviation effect, and Garwin and colleagues a 22-standard-deviation effect. (The probability of a 10-standard-deviation effect is 1.5×10^{-23}. In a lottery with a guaranteed winner that sells 10 million tickets, a buyer of 1 ticket has a better chance of winning the lottery three times in a row than of seeing a 10-standard-deviation effect.) As my former student Mark Corske remarked, "Four standard deviations is strong evidence, 13 standard deviations is absolute truth, and 22 standard deviations is the word of God."

Parity is not conserved in the weak interaction.

Overlooking Parity Nonconservation

The experimental results reported in the 1920s and 1930s that, at least in retrospect, showed the nonconservation of parity in weak interactions were performed by Richard Cox and his collaborators (1928) and by his student, Carl Chase (1929, 1930a, 1930b). The anomalous nature of these experimental results was fairly well known, although the exact nature of the anomaly was not clear. One thing is certain: the relationship of the results to the principle of parity conservation was not recognized or understood by any contemporary physicists, including the authors themselves.

These early experiments were part of the attempt to demonstrate the vector nature of electron waves. Louis De Broglie suggested in 1923 that just as light exhibits both particle and wave characteristics (light shows interference, a wave phenomenon, whereas in the photoelectric effect light behaves as a particle), so should those things that are normally considered particles, such as electrons or protons, exhibit wave characteristics. The wave nature of electrons was confirmed in 1927 in an experiment on the diffraction of electrons by crystals performed by Clinton Davisson and Lester Germer (1927). They had shown that in such experiments the electrons exhibited interference effects that were characteristic of waves. This idea of electron waves was then combined with the concept of electron spin by Charles Darwin (grandson of the Charles Darwin of evolutionary theory) to form the idea of a vector electron (1927). Cox and his collaborators thought that an experiment in which electrons were twice scattered from metal targets would provide evidence for the vector electron. In analogy with experiments on light and x-rays, the first scattering would polarize the electrons, resulting, for example, in more electrons with spin pointing in the positive x-direction than in the negative x-direction. The second scattering would detect that polarization. (If the electrons were polarized, the second scattering would result in an asymmetric result. For example, fewer electrons would be scattered in the forward direction than in the backward direction.)

Although the general nature of the effect to be observed in this experiment was known from the optical analogies, a detailed calculation of the effects expected was not carried out until the work of Nevill Mott in 1929. Mott calculated, on the basis of Paul Dirac's electron theory, that in the double scattering of electrons from heavy nuclei at large angles there would be a difference in the number of electrons scattered in the forward and backward directions (a 0°–180° asymmetry). If, on the other hand, the electron beam was initially longitudinally polarized, its spin either parallel to or opposite to the electron momentum, the number of electrons scattered at 90° and at 270° would be different, a left-right asymmetry. This latter possibility, which would indicate parity nonconservation, was not considered by Mott. The very existence of a longitudinal polarization for electrons from β-decay is also evidence for parity nonconservation. This is made clear by figure 2.1. In this case the spin is regarded as the spin of the electron itself, rather than that of the nucleus. Assuming that the electron spin is opposite to its momentum, a one-dimensional mirror reflection will reverse the spin direction, but the direction of the momentum will remain unchanged. The mirror image will have the spin in the same direction as the momentum, a clear difference. If the mirror image differs from the real object, parity is not conserved.

Cox and his colleagues described their experiment as follows: "In our experiment β-particles, twice scattered at right angles, enter a Geiger counter. The relative numbers entering are noted as the angle between the initial and final segments of the path is varied. . . . The angles at which most of the observations have been made are indicated as 270° and 90°. The difference between the configurations of the three segments of path at these two angles is the same as the *difference between right- and left-handed rectangular axes*" (Cox, McIlwraith, and Kurrelmeyer 1928; emphasis added). Their targets consisted of gold plugs, and a milligram of radium, a radioactive element, was used as the source of electrons. The scattered electrons were then detected by platinum-point Geiger counters. These Geiger counters had a short lifetime, and the points often had to be replaced. In addition, their behavior was inconsistent. Not all of the experimental runs showed an asymmetry. Cox and his collaborators stated, "It will be noted that of these results a large part indicate a marked asymmetry in the sense already mentioned. The rest show no asymmetry beyond the order of the probable error." The weighted average of their experimental results gave the ratio of the number of events at 90° to the number at 270° as 0.91 ± 0.01. This left-right asymmetry was a startling and unexpected result.

The experimenters then examined the possible sources of error in their experiment. They rejected all of these as unlikely and concluded,

"It should be remarked of several of these suggested explanations of the observations that their acceptance would offer greater difficulties in accounting for the discrepancies among the different results than would the acceptance of the hypothesis that we have here a true polarization due to the double scattering of asymmetrical electrons. This latter hypothesis seems the most tenable at the present time." The authors offered no theoretical explanation of their results, but they did suggest that the discrepancies in their results might be attributable to a velocity-dependent inefficiency of their Geiger counters. (Some of the counters used detected only the slower electrons, whereas the polarization effect was largest for faster electrons.)

Cox's experiments were continued by Carl Chase, a graduate student working under Cox's supervision. His early results, obtained with a Geiger counter as a detector, gave "no indication of polarization . . . of the kind suspected by Cox, McIlwraith, and Kurrelmeyer" (Chase 1929). By this time Mott's 1929 calculation had appeared, and Chase remarked that he had observed a small asymmetry between the counts at 0° and 180°, the forward-backward asymmetry predicted by Mott, but he attributed the effect to a difference in the paths that the electrons traveled in his apparatus.

Chase continued his work and found a substantial velocity dependence in the efficiency of the Geiger counters, as suggested earlier by Cox and his collaborators. Chase then redesigned and modified his experimental apparatus, using an electroscope rather than a Geiger counter to detect the scattered electrons, to avoid the difficulties involved with the use of those counters. His new experiment gave a ratio of 0.973 ± 0.004 (counts at 90°)/(counts at 270°). He concluded, "The following can be said of the of the present experiments: the asymmetry between the counts at 90° and 270° is always observed, which was in no sense true before. Not only every single run, but even all readings in every run, with few exceptions show the effect" (1930b). In this second experiment, Chase also obtained 0°–180° asymmetry of 0.985 ± 0.004. This time he believed that his result was not an artifact produced by his apparatus, and he did attribute it to a Mott scattering effect.

During the 1950s, after the initial experiments that demonstrated parity nonconservation, experiments on the double scattering of electrons were again performed with electrons from β-decay sources, an important point because only electrons from β-decay are initially longitudinally polarized. These later experiments obtained results quite similar to those of Cox and Chase and demonstrated the nonconservation of parity. As Cox remarked later, "It appears now in retrospect, that our experiments and those of Chase were the first to show evidence for parity nonconservation in weak interactions" (1973).

That was not, however, the reaction of the 1930s physics community. Although the results of Cox and Chase were occasionally mentioned as an anomaly in the literature on electron scattering, absolutely no recognition was accorded either by the authors or by anyone else to their significance for the question of parity nonconservation. Bernard Kurrelmeyer, a collaborator of Cox, stated, "As to our understanding of parity, it was nearly nil. Even the term had not been coined in 1927, and remember, this experiment was planned in 1925 and none of us were theoreticians" (personal communication, 1977). Cox, in discussing the reaction of the physics community, stated, "I should say that the experiments were widely ignored," and he added, "Our work was, prior to 1957, generally unaccepted, disbelieved, and poorly understood. Only by viewing it from the new theoretical framework and experimental observations of the late 50s could our results be comprehended" (1973).

There is an interesting and quite puzzling problem associated with the experimental results of Cox and of Chase. In 1959, Lee Grodzins recognized the relevance of those early results to the question of parity conservation. He concluded that these two experiments did indeed show a left-right (90°–270°) asymmetry and thus could have provided evidence for parity nonconservation. In a later publication, Grodzins pointed out that his earlier analysis was incorrect because both experiments had found fewer counts at 90° than at 270°, whereas contemporary theory predicts, and modern experiments demonstrate, more counts at 90°, and thus that both Chase and Cox had found an effect with the wrong sign. My own analysis, along with comparison between the results of experiments in the 1950s and those of Cox and Chase, confirmed that the sign of the asymmetry obtained by Chase and Cox was, in fact, wrong. Grodzins concluded that although the published sign of the asymmetry was incorrect, that Cox and Chase had carried out correct experiments: "It has long been my view that Chase and Cox did correct experiments, but that between the investigation and the write-up the sign got changed. . . . Did Cox mislabel his angles? Did he use a right-handed coordinate system instead of the left-handed one shown in his figure? If, as I suspect, he did make some such slip then the error would undoubtedly have been retained in subsequent papers. Such errors are neither difficult to make nor particularly rare. Many a researcher and at least one former historian of science have erred similarly" (1973).

Cox was initially unaware of Grodzins's later analysis. His own later recollections of the problem differ:

> I was quite surprised many years later when Lee Grodzins credited McIlwraith, Kurrelmeyer and myself with having been the first to observe parity violation. I was equally surprised; and naturally disappointed when he wrote in a later article that the asymmetry in the double scattering of β-rays, as described in our paper, was in the direction opposite to that predicted by the theory and that predicted by Yang and Lee. . . . I did not know, before the articles were printed, of the contradiction between the asymmetry predicted by the theory and that reported by McIlwraith, Kurrelmeyer, and myself, and by Chase. Grodzins in his article expressed the opinion that we (or I should say I, since I think our paper as published was mainly written by me) made a slip between the experimental observations and its published description. He supposes that the asymmetry we found was actually in the sense the theory predicts but that, in describing the experiment, I accidentally reversed it. At first sight, at least, this seems unlikely. But the alternative explanation, which assumes a persistent instrumental asymmetry, also seems unlikely when I consider how often we removed the Geiger counter to change electrodes (as was necessary in the early short-lived type of counter which McIlwraith, Kurrelmeyer and I used) and when I remember also other changes which Chase made in the very different equipment with which he replaced ours. I have thought about the matter off and on for a long time without coming to any conclusion either way. (Personal communication, 1977)

Although Cox was being cautious, his argument against a persistent instrumental asymmetry, in both his reminiscence above and the published paper, is convincing. In addition, the experiments of both Cox and Chase showed the velocity dependence of the polarization that is predicted by modern theory and that has been observed in later experiments. Despite the sign problem it does seem that those early experiments were the first to show evidence for parity nonconservation in weak interactions. (In a letter to me, Professor Cox indicated that he now agreed with my analysis that he had done a correct experiment but had made an error in the coordinate systems.)

Why were these experiments almost completely ignored by the physics community? The standard explanation is that the experiments were redone with electrons from heated metals, rather than from β-decay sources, which do not show the effect, so that they were dismissed: "As a cure the beta decay electrons were replaced with those from a hot filament, the effect disappeared and everybody was satisfied" (Frauenfelder and Henley 1975, 392). Although there is an element of truth to this explanation, it is by no means complete. No theoretical context was available at the time that suggested that these experiments were relevant to the question of parity nonconservation. Parity conservation itself had been suggested only in 1927. In addition, there were similar experiments, performed with the same type of apparatus, which, at the time, seemed to be far more important.

Cox's own recollections provide a useful starting point: "As to the reaction of other physicists to the experiment of McIlwraith, Kurrelmeyer, and myself, (and also to that of Chase on the same subject) I should say that the experiments were widely ignored. . . . Our reported results neither confirmed nor disproved any theory which was a subject of acute interest at the time" (personal communication, 1977).

At the time, no specific theoretical context existed into which to place these early experiments, in contrast to the situation in 1957 when the explicit theoretical predictions of Lee and Yang were published. Cox supports this view (1973): "During the nearly thirty years which passed between our experiments and those of Wu, Garwin, and Telegdi, many doubts were expressed about our observation. These doubts can be easily understood when one considers the theoretical models which prevailed before Lee and Yang. Our work was, prior to 1957, generally unaccepted, disbelieved, and poorly understood. Only by viewing it from the new theoretical framework and experimental observations of the late 50s, could our results be comprehended."

It is understandable that these early experiments were overlooked because of the lack of theoretical predictions. What is still puzzling is why the perceived anomaly in the results did not act as a stimulus for further work, both experimental and theoretical, in the same way as the θ-τ puzzle did in the 1950s and why these results were ultimately ignored. I suggest that they became lost in the struggle of scientists to corroborate the predictions of Mott that there should be forward-backward (0°–180°) asymmetry in the double scattering of electrons (1929). That result, which tested an important, well-supported, and accepted theory, seemed to be far more important. Mott's calculation was based on Dirac's relativistic electron theory, so that any apparent refutation of Mott's theory also cast doubt on Dirac's theory, which was strongly believed on other grounds. (Dirac's was the only theory at the time that predicted the existence of the positron, a positively charged electron. The observation of the positron in 1932 by Carl Anderson provided strong support for Dirac's theory [Anderson 1933].)

Experiments on the double scattering of electrons began in the mid-1920s, and the general problem of electron scattering from nuclei, as well as the discrepancy between the experimental results and the specific predictions by Mott, were of concern until the 1940s. Difficulties with the consistency of experimental results and subtle and unforeseen effects in electron scattering were present throughout.

With the exception of the result of Cox and his collaborators, none of the experiments performed before 1929 showed any evidence of electron polarization. Change came in 1929 with the publication of Mott's theoretical calculation of the double scattering of electrons. Mott's calculation was based directly on Dirac's relativistic electron theory and made specific theoretical predictions concerning the asymmetry to be observed in the double-scattering experiment. Mott predicted that there would be a forward-backward (0°–180°) asymmetry in the double scattering of initially unpolarized electrons. He specified the specific conditions under which this asymmetry should be observed, namely, single, large-angle scattering from nuclei with a large charge. In later work he also provided precise numerical values expected for the asymmetry. But he noted that his theory did not predict any asymmetry between the left and right directions. "It was in this plane [left-right] that asymmetry was looked for by Cox and Kurrelmeyer, and the asymmetry found by them must be due to some other cause" (1929). Mott was not questioning the correctness of the experimental results of Cox and colleagues and Chase, he was merely noting that his theory did not explain them.

Subsequent experimental work in the 1930s took on a different character following Mott's researches, because there were then explicit theoretical predictions, based on an accepted theory, with which to compare the experimental results. The experimental situation was confused at best, but no attempts were made to replicate the Cox-Chase results. All of the experiments were designed to test Mott's theory and to search for a forward-backward asymmetry. Some experimenters found the predicted results, others did similar experiments and obtained null results, and some experimenters found positive results at one time but not at others. In general, the trend in experimental results was in disagreement with Mott's calculation. This discrepancy between theory and experiment led not only to further experimental work, but also to many unsuccessful attempts by theoretical physicists to provide reasons for the absence of the predicted polarization effects.

By far the most positive evidence in favor of Mott's calculation was provided by Emil Rupp. In a series of papers during the early 1930s, Rupp reported results in general agreement with those predicted by Mott (Rupp 1929, 1930, 1931, 1932a, 1932b, 1932c, 1934; Rupp and Szilard 1931). These results, which differed from the primarily negative results found by almost everyone else, served to confuse the issue of whether the polarization effects predicted by Mott had been observed. It was soon revealed that Rupp's results were fraudulent. In 1935, Rupp published a formal withdrawal of several of his results. This paper contained a note from a psychiatrist stating that for the past several years Rupp had suffered from a mental illness and could not distinguish between fantasy and reality. There are reports that after Rupp's withdrawal was published, his locked laboratory was revealed to contain either no equipment for performing electron-scattering experiments or only apparatus for forging data. The anecdotes differ. (For more details of Rupp's career, see French 1999.)

In 1937 H. Richter published what he regarded as the definitive experiment on the double scattering of electrons. He claimed to have satisfied the conditions of Mott's calculation exactly and found no effect. He concluded, "Despite all the favorable conditions of the experiment, however, no sign of the Mott effect could be observed. *With this experimental finding, Mott's theory of the double scattering of electrons from the atomic nucleus can no longer be maintained.*"

There was a definite discrepancy between Mott's theory and the experimental results, and that discrepancy continued despite various theoretical attempts to remove it. As Morris Rose and Hans Bethe concluded (1939), "Unfortunately, none of the effects considered produces any appreciable depolarization of the electrons and the discrepancy between theory and experiment remains—perhaps more glaring than before."

Ironically, the solution was provided by the experimental work of Cox, Chase, and their collaborators in the early 1940s. They found that an experimental artifact had precluded the observation of the predicted effects. This became known as the reflection-transmission effect. In a double-scattering experiment, two different types of experimental apparatus are used: one in which the electrons pass through the thin-foil targets, a transmission experiment, and a second in which the electrons are scattered from the front surface of the foil, a reflection experiment. To minimize the effects of multiple scattering, an important background effect, all of the experiments performed in the 1930s were reflection experiments.

The work of Cox and collaborators in the 1940s showed that in such reflection experiments "plural scattering," in which a large-angle scattering is made up of a few smaller-angle scatterings, will mask the effect of single scattering. Because the plural-scattering electrons are unpolarized, the effect predicted by Mott will not be observed. In a transmission experiment, plural scattering is far less important and the predicted effect can be seen. When this was realized, the experiments were redesigned and the discrepancy between theory and experiment removed. At this point, however, not even Cox and his collaborators remembered their earlier left-right asymmetry result, and the double-scattering experiments on that asymmetry were not repeated until the 1950s, after the discovery of parity nonconservation.

Scientists are not omniscient. They do not always realize all of the implications of either experimental results or theoretical calculations. Clearly, the experiment of Cox, McIlwraith, and Kurrelmeyer and those of Chase show, at least in retrospect, the nonconservation of parity. In this episode a strongly believed scientific hypothesis, parity conservation, was overthrown, a decision based on overwhelming experimental evidence. Things are not always so clear and unambiguous in the practice of science.

Experiments that, in retrospect, showed parity nonconservation were not understood by either the experimenters themselves or anyone else in the physics community. At least part of the reason for the failure to recognize the importance of the experiments of Cox and his collaborators and of Chase was the lack of a theoretical context in which to place the work. Such a context existed in 1956 because of the work of Lee and Yang. Cox and his collaborators did come tantalizingly close to recognizing the implications of their work: "The difference between the configurations of the three segments of path at these two angles is the same as the *difference between right- and left-handed rectangular axes*" (Cox, McIlwraith, and Kurrelmeyer 1928).

In the difficult investigation of Mott scattering of electrons, which seemed to be a more important problem at the time, these experimental results were also neglected. The fallibility of science is quite clear in this episode. Experiments gave conflicting answers about Mott scattering during the 1930s, but ultimately a consensus was reached that experiment disagreed with theory. The failure of the experiments to agree with Mott's predictions cast doubt on Dirac's theory, which had other substantial evidential support. This episode also illustrates one way in which the physics community reacts to a seemingly clear discrepancy between experimental results and a well-corroborated theory. Dirac's relativistic electron theory, on which Mott's calculation had been based, was not rejected or regarded as refuted, even after many repetitions had seemed to establish the discrepancy beyond any doubt. The tenacity and perseverance of the physics community led to many repetitions of the experiment, under similar and under slightly different conditions. Various theoretical suggestions were made to try to solve the problem, all of which were unsuccessful. The discrepancy was finally resolved by an experimental demonstration, followed by a theoretical explanation, of why the earlier experimental results were wrong.

Experimental evidence and reasoned and critical discussion played an essential role in this episode. It was good—albeit fallible—science.

(3)

The Meselson-Stahl Experiment

"THE MOST BEAUTIFUL EXPERIMENT IN BIOLOGY"

A renowned experiment carried out by two postdoctoral fellows decided among three competing mechanisms for the replication of DNA, the molecule now believed to be responsible for heredity. This experiment strongly supported one proposed mechanism and argued against the other two.

In 1953 Francis Crick and James Watson proposed a three-dimensional structure for deoxyribonucleic acid (DNA; Watson and Crick 1953a, 1953b). Their proposed structure consisted of two polynucleotide chains helically wound about a common axis—the famous double helix. The chains were bound together by combinations of four nitrogen bases: adenine, thymine, cytosine, and guanine. The nitrogen base on one chain was hydrogen-bonded to the base at the same level in the other chain. Because of structural requirements, only the base pairs adenine-thymine and cytosine-guanine were allowed. This explained the previously observed regularity, known as Chargaff's rules, that the

For the history of this complex episode, see Holmes (2001).

amount of adenine contained in the DNA of any species is approximately equal to the amount of thymine. The same is true for cytosine and guanine. Each chain is thus complementary to the other. If there is an adenine base at a location in one chain, there is a thymine base at the same location on the other chain, and vice versa. The same applies to cytosine and guanine. The order of the bases along a chain is not, however, restricted in any way, and it is the precise sequence of bases that carries the genetic information. "This brilliant accomplishment ranks as one of the most significant in the history of biology because it led the way to an understanding of gene function in molecular terms" (Stryer 1975, 563).

The significance of the proposed structure was not lost on Watson and Crick when they made their suggestion. They remarked, "It has not escaped our notice that the specific pairing we have postulated immediately suggests a possible copying mechanism for the genetic material" (1953a).

If DNA was to play this crucial role in genetics, then there must be a mechanism for the replication of the molecule. Within a short time following the Watson-Crick suggestion, three different mechanisms for the replication of the DNA molecule were proposed (Delbruck and Stent 1957). These are illustrated in figure 3.1 The first, proposed by Gunther Stent and known as conservative replication, suggested that each of the two strands of the parent DNA molecule was replicated in new material. This would yield a first generation consisting of the original parent DNA molecule and one newly synthesized DNA molecule. The second generation would consist of the parental DNA and three new DNAs.

The second mechanism, known as semiconservative replication, proposed that each strand of the parental DNA acted as a template for a second, newly synthesized complementary strand, which would then combine with the original strand to form a DNA molecule. This was proposed by Watson and Crick (1953b). The first generation would consist of two hybrid molecules, each containing one strand of parental DNA and one newly synthesized strand. The second generation would be two hybrid molecules and two totally new DNAs. The third mechanism, proposed by Max Delbruck, was dispersive replication, in which

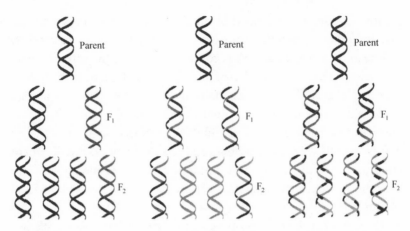

FIGURE 3.1. Possible mechanisms for DNA replication. *Left*, conservative replication. Each of the two strands of the parent DNA is replicated to yield the unchanged parent DNA and one newly synthesized DNA. The second generation consists of one parent DNA and three new DNAs. *Center*, semiconservative replication. Each first-generation DNA molecule contains one strand of the parent DNA and one newly synthesized strand. The second generation consists of two hybrid DNAs and two new DNAs. *Right*, dispersive replication. The parent chains break at intervals, and the parental segments combine with new segments to form the daughter chains. The darker segments are parental DNA, and the lighter segments are newly synthesized DNA. From Lehninger (1975).

the parental DNA chains would break at intervals and the parental segments would combine with new segments to form the daughter strands.

Matthew Meselson and Franklin Stahl, in what is called by John Cairns "the most beautiful experiment in biology" (quoted in Holmes 2001, 329), proposed to solve the problem of the DNA replication mechanism. They described their proposed method: "We anticipated that a label which imparts to the DNA molecule an increased density might permit an analysis of this distribution by sedimentation techniques. To this end a method was developed for the detection of small density differences among macromolecules. By use of this method, we have observed the distribution of the heavy nitrogen isotope ^{15}N among molecules of DNA following the transfer of a uniformly ^{15}N-labeled, exponentially growing bacterial population to a growth medium containing the ordinary nitrogen isotope ^{14}N" (Meselson and Stahl 1958).

The experiment is described schematically in figure 3.2. Meselson and Stahl placed a sample of DNA in a solution of cesium chloride. As the sample is rotated at high speed, the denser material travels further away from the axis of rotation than does the less-dense material. This results in a solution of cesium chloride with density that increases in the direction away from the axis of rotation. The DNA reaches equilibrium at the position where its density equals that of the solution. Meselson and Stahl grew *Escherichia coli* bacteria in a medium that contained ammonium chloride (NH_4Cl) as the sole source of nitrogen. They did this for media that contained either ^{14}N, ordinary nitrogen, or ^{15}N, a heavier isotope. By destroying the cell membranes, they could obtain samples of DNA that contained either ^{14}N or ^{15}N. They first showed that they could indeed separate the two different mass molecules of DNA by centrifugation (figure 3.3). The separation of the two types of DNA is clear in both the photograph obtained by absorbing ultraviolet light and in the graph showing the intensity of the signal, obtained with a densitometer. (Both samples contained approximately the same number of cells.) In addition, the separation between the two peaks suggested that they would be able to distinguish an intermediate band composed of hybrid DNA from the heavy and light bands.

Meselson and Stahl then produced a sample of *E. coli* containing only ^{15}N by growing it in a medium containing only ammonium chloride with ^{15}N ($^{15}NH_4Cl$) for 14 generations. They then abruptly changed the medium to ^{14}N by adding a 10-fold excess of $^{14}NH_4Cl$. Samples were taken just before the addition of ^{14}N and at intervals afterward for several generations. The cell membranes were broken to release the DNA into the solution, and the samples were centrifuged and ultraviolet absorption photographs taken. In addition, the photographs were scanned with a recording densitometer. The results are shown in figure 3.4, with both the photographs and the densitometer traces. The figure shows that the researchers start with only heavy (fully labeled) DNA. As time proceeds more and more half-labeled DNA is present, until at one generation there is only half-labeled DNA. "Subsequently only half labeled DNA and completely unlabeled DNA are found. When two generation times have elapsed after the addition of ^{14}N half-labeled and unlabeled DNA are present in equal amounts" (Meselson and Stahl 1958).

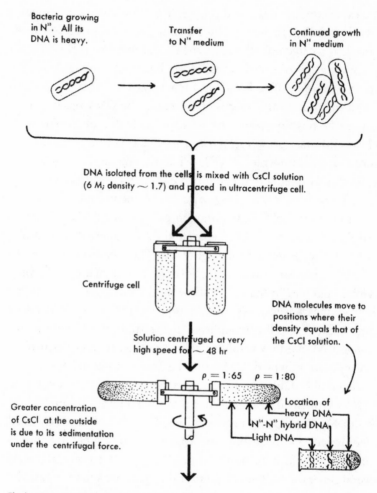

Bacteria growing in N[15]. All its DNA is heavy.

Transfer to N[14] medium

Continued growth in N[14] medium

DNA isolated from the cells is mixed with CsCl solution (6 M; density ∼ 1.7) and placed in ultracentrifuge cell.

Centrifuge cell

Solution centrifuged at very high speed for ∼ 48 hr

DNA molecules move to positions where their density equals that of the CsCl solution.

$\rho = 1:65$ $\rho = 1:80$

Greater concentration of CsCl at the outside is due to its sedimentation under the centrifugal force.

Location of
heavy DNA
N[14]-N[15] hybrid DNA
Light DNA

The location of DNA molecules within the centrifuge cell can be determined by ultraviolet optics. DNA solutions absorb strongly at 2600 A.

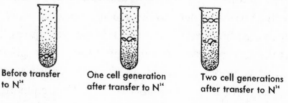

Before transfer to N[14]

One cell generation after transfer to N[14]

Two cell generations after transfer to N[14]

FIGURE 3.2. Schematic representation of the Meselson-Stahl experiment. From Watson (1965).

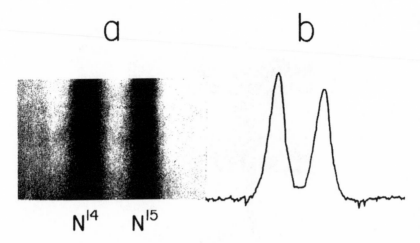

a   b

N^{14} N^{15}

FIGURE 3.3. The separation of ^{14}N DNA from ^{15}N DNA by centrifugation: a, ^{14}N DNA, ^{15}N DNA; b, densitometer trace. From Meselson and Stahl (1958).

(This is exactly what the semiconservative replication mechanism predicts.) By four generations the sample consists almost entirely of unlabeled DNA. The conclusion that the DNA in the intermediate-density band was half-labeled was tested by examining a sample containing equal amounts of generations 0 and 1.9. If the semiconservative mechanism is correct, then generation 1.9 should have approximately equal amounts of unlabeled and half-labeled DNA, whereas generation 0 would contain only fully labeled DNA. As shown, there are three clear density bands, and Meselson and Stahl found that the intermediate band was centered at (50 ± 2) percent of the difference between the ^{14}N and ^{15}N bands, shown in the bottom photograph (generations 0 and 4.1). This is precisely what one would expect if the DNA was half-labeled.

Meselson and Stahl stated their results as follows:

1. *The nitrogen of DNA is divided equally between two subunits which remain intact through many generations.*

The observation that parental nitrogen is found only in half-labeled molecules at all times after the passage of one generation time demonstrates the existence in each DNA molecule of two subunits containing equal amounts of nitrogen. The finding that at the

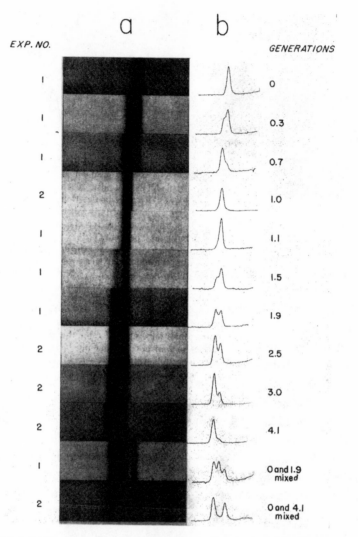

FIGURE 3.4. DNA replication: *a*, ultraviolet absorption photographs showing DNA bands from centrifugation of DNA from *Escherichia coli* sampled at various times after the addition of an excess of ^{14}N substrates to a growing ^{15}N culture; *b*, densitometer traces of the photographs. The initial sample is all heavy (^{15}N DNA). As time proceeds a second intermediate band begins to appear until at one generation all of the sample is of intermediate mass (hybrid DNA). At longer times a band of light DNA appears, until at 4.1 generations the sample is almost all lighter DNA. This is exactly what is predicted by the Watson-Crick semiconservative mechanism. From Meselson and Stahl (1958).

second generation half-labeled and unlabeled molecules are found in equal amounts shows that the number of surviving parental sub-units is twice the number of parent molecules initially present. That is, the subunits are conserved.

2. *Following replication, each daughter molecule has received one parental subunit.*

The finding that all DNA molecules are half-labeled one generation time after the addition of ^{14}N shows that each daughter molecule receives one parental subunit. If the parental subunits had segregated in any other way among the daughter molecules, there would have been found at the first generation some fully labeled and some unlabeled DNA molecules, representing those daughters which received two or no parental subunits, respectively. (Meselson and Stahl 1958)

Meselson and Stahl also noted implications of their work for deciding among the proposed mechanisms for DNA replication. In a section labeled "The Watson-Crick Model" they wrote,

This [the structure of the DNA molecule] suggested to Watson and Crick a definite and structurally plausible hypothesis for the duplication of the DNA molecule. According to this idea, the two chains separate, exposing the hydrogen-bonding sites of the bases. Then, in accord with base-pairing restrictions, each chain serves as a template for the synthesis of its complement. Accordingly, each daughter molecule contains one of the parental chains paired with a newly synthesized chain.

The results of the present experiment are in exact accord with the expectations of the Watson-Crick model for DNA replication. (Emphasis added)

Although it wasn't stated, the results also disagreed with the predictions of the conservative model of DNA replication, which predicted that in the first generation there would be one fully labeled molecule of DNA and one unlabeled molecule.

One question that still remained was what were the subunits of DNA that Meselson and Stahl had found. They remarked, "However, it must be emphasized that it has not been shown that the molecular subunits found in the present experiment are single polynucleotide chains or even that the DNA molecules studied here correspond to single DNA molecules possessing the structure proposed by Watson and Crick" (Meselson and Stahl 1958).

Meselson and Stahl obtained some information on that question from experiments on heated DNA. This heating dissociated the subunits of the DNA molecule. They found that the mass of the subunits, for both ^{14}N DNA and ^{15}N DNA, was approximately half that of the DNA molecule. This suggested, but did not prove, that the subunits of DNA were the entire single-polynucleotide chains of the Watson-Crick model of DNA structure. (This was demonstrated a few years later in the work of John Cairns [1961] and others.) It also showed that the dispersive replication mechanism proposed by Delbruck, which had smaller subunits, was incorrect. "Since the apparent molecular weight of the subunits so obtained is found to be close to half that of the intact molecule, it may be further concluded that the subunits of the DNA molecule which are conserved at duplication are single, continuous structures. The scheme for DNA duplication proposed by Delbruck is thereby ruled out" (Meselson and Stahl 1958).

Meselson and Stahl rather modestly concluded,

> The structure for DNA proposed by Watson and Crick brought forth a number of proposals as to how such a molecule might replicate. These proposals make specific predictions concerning the distribution of parental atoms among progeny molecules. The results presented here give a detailed answer to the question of this distribution and simultaneously direct our attention to other problems whose solution must be the next step in progress toward a complete understanding of the molecular basis of DNA duplication. What are the molecular structures of the subunits of *E. Coli* DNA which are passed on intact to each daughter molecule? What is the relationship of these subunits to each other in a DNA molecule? What is the mechanism of the synthesis and dissociation of the subunits in vivo?

Historian of science Larry Holmes, author of the definitive history of the Meselson-Stahl experiment, wrote,

> Both Meselson and Stahl were personally convinced that the experiment had proven the position Watson and Crick had taken in 1953 to be right: that what they had observed was the separation and conservation of the two polynucleotide strands of the DNA double helix.

Mention of a molecular weight, a frictional coefficient, and denaturation behavior consistent with those of DNA molecules was intended to suggest cautiously the likelihood that the "unit" of the experiment were such molecules. Meselson's resistance to the temptation to state his belief, even in an informal letter, that they actually were "WC duplexes" was due to a concern not to draw conclusions that went beyond what his data said. (2001, 329)

Even before the paper was published, the results were known to many of those working in the field by means of letters, phone calls, seminars, and visits. (These days papers are often available on the Internet or can be transmitted electronically.) They were received enthusiastically. Prior to the publication of the results, Christian Anfisen, who was writing a book titled *The Molecular Basis of Evolution*, requested photographs of the results for inclusion in his text. Textbooks convey what is accepted as knowledge in a field at a given time, so Anfisen's request shows that the Meselson-Stahl results were already being accepted as correct.

There were other, more informal measures of that acceptance. Meselson visited Stent at Berkeley and showed him the photographs. When Stent saw them he realized that DNA did not replicate conservatively and complained cheerfully about the "devilish, hellish experiment," whose results were not open to challenge. At a seminar given by Meselson during his visit, Stent introduced him as "the Mozart of molecular biology" (Holmes 2001, 325). Stent still had some reservations. In a letter to Delbruck, at the time, he wrote, "Things *do* look good for the WC [Watson-Crick] mechanism, I must admit (very generous of me, isn't it?), but I *still* can't see how the intimate connection between replication and genetic exchange can be accounted for from this point of view" (quoted in Holmes 2001, 325).

Although Delbruck's dispersive mechanism had fared no better than Stent's under the results of the Meselson-Stahl experiment, Delbruck was even more positive. In a letter to a friend he wrote, "Meselson is making earth-shaking discoveries in the replication of DNA; and every afternoon he and Simsheimer and I have been having endless discussions (with tea) about this" (quoted in Holmes 2001, 326).

Meselson also wrote to Watson telling him of the results. Before he and Stahl had performed the experiment they had speculated about its possible outcomes and written verses to express them. Meselson included one in his letter to Watson, although it did not, in fact, apply to the results.

> Now N^{15} by heavy trickery
> Ends the sway of Watson-Crickery
> But now we have WC with a mighty vengeance . . . or else a diabolical camouflage. (Quoted in Holmes 2001, 326–27)

Although Watson probably did not write a detailed answer to Meselson's letter, he did call others he thought would be interested in the results. One of the recipients, Cyrus Levinthal at the Massachusetts Institute of Technology, wrote to Meselson, "Again my congratulations. The experiment is extremely beautiful" (quoted in Holmes 2001, 327).

Meselson also sent a mimeographed copy of his paper to Maurice Wilkins, who would later share the Nobel Prize with Watson and Crick for the discovery of the structure of DNA. Wilkins replied, "Thank you very much for your letter and the mss [manuscript] describing your elegant and definitive experiments. . . . As a result of your experiments I personally begin to feel real confidence in the Watson and Crick duplication hypothesis" (quoted in Holmes 2001, 387).

Meselson also made a composite photograph, similar to figure 3.4, which showed the results, and showed it to various people. Linus Pauling's reaction was that Meselson and Stahl had carried out "an extraordinary experiment" (quoted in Holmes 2001, 337). Meselson also showed the results to Richard Feynman, the theoretical physicist. Feynman found them to be "tremendously exciting and very important" (quoted in Holmes 2001, 337). As Holmes recounts (2001, 337), "He urged Meselson to state his claim that the experiment confirmed the mode of replication predicted by the double helix, but even the enthusiasm of this brilliant physicist could not induce Meselson to abandon his reserve about the nature of the subunits."

The anecdotal evidence seems clear. The Meselson-Stahl experiment had provided strong evidence for semiconservative replication and against the other DNA replication mechanisms.

Textbooks, which tell us what is accepted as knowledge, confirm the personal judgments. Watson's 1965 text, *Molecular Biology of the Gene*, contains a section titled "Solid Evidence in Favor of DNA Strand Separation," in which he describes the Meselson-Stahl experiment (figure

3.2 is from Watson's book), although he makes no mention of Mesel-son or Stahl. Typical of texts written in the 1970s and 1980s are Albert Lehninger's *Biochemistry* (1975) and Lubert Stryer's *Biochemistry* (1975). Lehninger remarks (p. 892), "Ingeniously contrived experiments carried out by M.S. Meselson and F.W. Stahl in 1957 conclusively proved that in intact living *E. coli* cells DNA is replicated in the semiconservative manner postulated by Watson and Crick." He further refers to the work as a "crucial experiment." Stryer has a section titled "DNA Replication is Semiconservative" and refers to the Meselson-Stahl experiment as a "critical test" (pp. 568–70). He does, however, cite the cautious conclu-sion of the Meselson-Stahl paper.

Holmes summarized the effect of the experiment (2001, 436): "The Meselson-Stahl experiment marks, as prominently as any other single event, the transition between an era in which the reality of the replica-tion mechanism suggested by the double helix was contested and one in which the assumption of its fundamental correctness sustained in-vestigations of the details of the process. That search, continuing to the present, has revealed a mechanism far more complex than could be imagined by those who attempted, during the first years of the double helix, to picture how its strands might come apart and serve as templates of further complementary strands."

Still, there is no reason to question that the Meselson-Stahl experi-ment is indeed an extremely important, if not crucial, experiment. As John Cairns stated,

In a sense everything comes together. First of all, the experiment was needed at the time. Secondly . . . the situation was such that that was the result. Suppose [that DNA replicated] in tiny pieces as Delbruck and Stent said, then the result would not have been so beautiful. . . . But I don't think Meselson and Stahl get any credit for that sector of the beauty, nature gets the credit for that.

Then there was the beauty of the way they did it, because they did it with the utmost precision. Lastly, there was the fact that they knew every implication of [what] they found. . . . Matt and Frank wrung every bit out of [the result], as elegantly as possible. So it has elegance on every count, and the fact that it has it on every count is itself an ele-gance. (Quoted in Holmes 2001, 249)

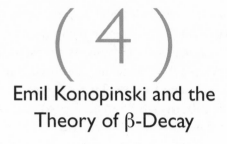

Emil Konopinski and the
Theory of β-Decay

George Levine, a sympathetic and critical commentator on science, once asked (1987, 13), "What if important scientific discoveries were often made because the scientist *wanted* something to be true rather than because he or she had evidence to prove it true?" Emil Konopinski was a scientist who resisted the temptation to bend the facts to his theory. In the 1930s, he was a coauthor of an alternative to Enrico Fermi's theory of β-decay. Although Konopinski's theory initially seemed to be supported by the existing experimental evidence, further work provided considerable evidence against his theory. However much he wanted the theory to be correct and to make a judgment against the evidence, he did not.

Experiment decided the issue between the two competing theories of β-decay. This history illustrates both the fallibility of science and the ability of science to overcome errors. The early experimental results on β-decay, which suggested the need for an alternative to Fermi's

For a more detailed history of this episode, see Franklin (1990, chapter 1).

theory, were incorrect. The electrons emitted by the radioactive elements lost energy in escaping from the thick sources used, giving rise to an excess of low-energy electrons. There was, in addition, an incorrect theory-experiment comparison. A correct, but inappropriate, theory was compared to the experimental results. It was only when both of these errors were corrected that a valid experiment-theory comparison was made. Then Fermi's theory was supported.

In the β-decay process, an atomic nucleus emits an electron, simultaneously transforming itself into a different kind of nucleus. In the early 20th century it was thought that the final state of β-decay involved only two bodies (the daughter nucleus and the electron). The laws of conservation of energy and conservation of momentum require that the electron emitted in such a two-body decay be monoenergetic. For each radioactive element, all the electrons emitted have the same energy. Considerable experimental work on β-decay in this period culminated in the demonstration by Charles Ellis and William Wooster that the energy spectrum of the emitted electrons was continuous (1927). The electrons were emitted with all energies from zero up to a maximum, which depended on the particular element. Such a continuous spectrum was incompatible with a two-body process. The conservation laws were threatened. In 1930, when Wolfgang Pauli proposed that a third particle, one that was electrically neutral, had a very small mass, and had spin ½, was also emitted in β-decay. This solved the problem and saved the conservation laws because in a three-body process (i.e., neutron → proton + electron + third particle), the electron is not required to be monoenergetic but can have a continuous energy spectrum.

Determining the Correct Theory of β-Decay

Enrico Fermi named the proposed new particle the neutrino, little neutral one, and in 1934 incorporated it into a quantitative theory of β-decay (1934a; 1934b). (Fermi was one of the great physicists of the 20th century. Not only did he do important work in theoretical physics, including his theory of β-decay, but his experimental work was of the highest quality. In 1938 he was awarded the Nobel Prize for his demonstration of new radioactive elements produced by neutron irradiation and for his work on neutron reactions. During World War II he led the team that produced the first nuclear chain reaction.) Fermi assumed that the neutrino existed, that the atomic nucleus contained only protons and neutrons, and that the electron and the neutrino were created at the moment of decay. Fermi's theory gave a reasonable, although not exact, fit to the energy spectrum of electrons emitted in β-decay.

Although, as Konopinski and Uhlenbeck (1935) pointed out, Fermi's theory was "in general agreement with the experimental facts," more detailed examination of the decay energy spectra showed discrepancies. Fermi's theory predicted too few low-energy electrons and an average decay energy that was too high. Konopinski and Uhlenbeck cited as evidence the energy spectrum of ^{30}P obtained by Ellis and Henderson (1934) and that of RaE (^{210}Bi) measured by Sargent (1932, 1933; figure 4.1). It is clear that the curves labeled F • S (Fermi theory) are not a good fit to the observed spectra.

In 1935 Konopinski and Uhlenbeck proposed a modification of Fermi's theory that eliminated the discrepancy and that predicted more low-energy electrons. In figure 4.1, the curves labeled modified theor. (Konopinski-Uhlenbeck theory) fit the observed spectra far better than do the curves for the Fermi theory.

The Konopinski-Uhlenbeck theory was almost immediately accepted by the physics community as superior to Fermi's theory and as the preferred theory of β-decay. In a 1936 review article on nuclear physics, which remained a standard reference and was used as a student text into the 1950s, Hans Bethe and Robert Bacher, after surveying the experimental evidence, remarked, "We shall therefore accept the K-U theory as the basis for future discussions." This article was often referred to as the Bethe bible.

The K-U theory received substantial additional support from the results of the cloud-chamber experiment of F. Kurie, J. Richardson, and H. Paxton (1936). They found that the observed β-decay spectra of several elements all fit the K-U theory better than did the original Fermi theory. It was in this paper that the Kurie plot, which made comparison between the two theories far easier, made its first appearance. The Kurie plot was a graph of a particular mathematical function involv-

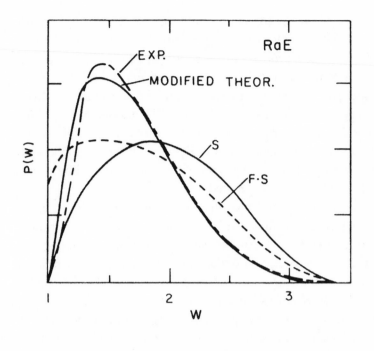

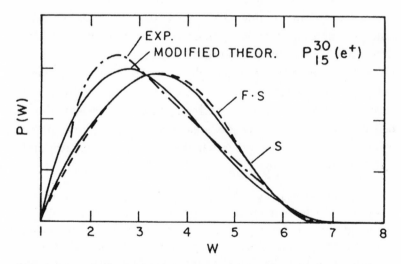

FIGURE 4.1. The energy spectra for the β-decay of RaE and ^{30}P, respectively. The curve labeled EXP is the experimental result, F • S is the Fermi theory, and Modified Theor. is the Konopinski-Uhlenbeck theory. The K-U theory is a better fit to the observed spectrum than is Fermi's theory. From Konopinski and Uhlenbeck (1935).

ing the electron energy spectrum plotted against the electron energy that gave different results for the K-U theory and for the Fermi theory. It had the nice visual property that the Kurie plot for whichever theory was correct would be a straight line. If the theory did not fit the observed spectrum, then the Kurie plot for that theory would be a curve. The Kurie plot obtained by Kurie and his collaborators for ^{32}P (phosphorus) is shown in figure 4.2. "The (black) points marked 'K-U' modification should fall as they do on a straight line. If the Fermi theory is being followed the (white) points should follow a straight line as they clearly do not," the authors pointed out.

Interestingly, Kurie and his collaborators had originally obtained results that were in agreement with the Fermi theory. Their experimental apparatus used a cloud chamber placed in a magnetic field. The electron tracks produced in the cloud chamber had a radius of curvature proportional to the momentum of the electrons. This also determined the energy of the electrons. Kurie and his collaborators attributed their incorrect initial result to the preferential elimination of low-energy decay electrons by one of their selection criteria, which eliminated events in which the electron tracks in the cloud chamber showed a visible deflection. An accurate measurement of the electron momentum (energy) cannot be obtained with a track that contains a large deflection. Not only does the momentum change, but the deflection makes fitting the observed track to a single track with constant momentum inaccurate. Low-energy electrons are scattered more frequently than high-energy electrons and will therefore have more tracks with visible deflections. The scattering was greatly reduced by filling the cloud chamber with hydrogen rather than the original oxygen. Kurie and his collaborators wrote (1936),

> Last spring we examined the Fermi theory to see if it predicted the shape of the distributions we were getting and reported favorably on the agreement between the two . . . with the reservation that we did not feel that our experiments were good enough satisfactorily to test a theory. At that time we were using oxygen as the gas in the chamber as is usual in β-ray work. In measuring the curves we had adopted the rule that all tracks with visible deflections in them were to be rejected. That this was distorting the shape of the distribution we knew because

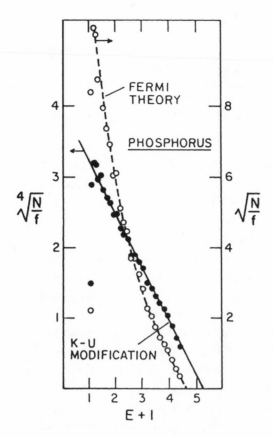

FIGURE 4.2. The Kurie plot for the decay of ^{32}P. "The (black) points marked 'K-U' modification should fall as they do on a straight line. If the Fermi theory is being followed the (white) points should follow a straight line as they clearly do not." From Kurie, Richardson and Paxton (1936).

we were being forced to discard many more of the low-energy tracks than the high energy ones. This distortion can be reduced to a very great extent by photographing the β-tracks in hydrogen instead of oxygen. The scattering is thus reduced by a factor of 64. [The scattering of an electron by a nucleus is proportional to Z^2, where Z is the charge on the nucleus. Thus electron scattering from oxygen, $Z = 8$, is 64 times larger than that from hydrogen, $Z = 1$.]

We found with the hydrogen filled chamber that the distribution curves were more skew than they had appeared with the oxygen filled chamber. This is not surprising: our criterion of selection had been forcing us to discard as unmeasurable a large number of low-energy tracks. The number discarded increased as the energy of the track decreased. The apparent concordance between our early data and the Fermi theory was entirely traceable to this because the Fermi distribution is very nearly symmetrical so that when the number of low-energy tracks was measured this apparent asymmetry in the experimental distributions was lost.

Similar problems affected other cloud-chamber experiments. Paxton found a different solution: he measured all tracks of sufficient length. "Because β-ray scattering becomes increasingly serious as the energy decreases, all tracks of sufficient length were measured as well as possible in spite of bad curvature changes, in order to prevent distribution distortion from selection criteria" (1937). Measuring a track with such a curvature change will usually result in an incorrect value of the momentum or energy of the particle. In addition, it will increase the uncertainty of that determination.

Additional support for the K-U theory came from several further measurements on the spectrum of radium E, a product of the radioactive decay of radium that is an isotope of bismuth (^{210}Bi), but the support provided for the K-U theory by all the available evidence was not unequivocal. Richardson pointed out that scattering and energy loss by electrons leaving the radioactive source could distort the energy spectrum, particularly at the low-energy end. "The failure of theory to explain the continuous spectrum makes it of interest to obtain all possible experimental information, and although much is now known about the high energy part of the curve, the low energy region has remained obscure owing to certain experimental difficulties. The chief of these has been the contamination of the low energy end of the curve by rays reflected with unknown energy from the material on which the radioactive body was deposited" (1934).

There were other uncertainties in the measurement of the RaE decay spectrum. In 1937 J. S. O'Conor remarked, "Since the original work of Schmidt in 1907 more than a score of workers have made measurements on the beta-ray spectrum of radium E with none too concordant results" (1937). He cited 27 different measurements of the high-energy end-point energy, for which the largest and smallest values differed by more than a factor of two. By 1941, however, a consensus was reached, and as A. Townsend stated, "The features of the β-ray spectrum of RaE are now known with reasonable precision." The future would be different. The spectrum of RaE would remain a problem until the 1950s, when it was finally solved.

The Kurie paper also discussed a problem faced by the K-U theory. The maximum decay energy extrapolated from the straight-line graph of the Kurie plot seemed to be higher than the value obtained visually from the energy spectrum. This discrepancy between the measured maximum electron energy and that extrapolated from the K-U theory persisted and became more severe as experiments became more precise. In 1937 Milton Livingston and Hans Bethe remarked, "Kurie, Richardson, and Paxton have indicated how the K-U theory can be used to obtain a value for the theoretical energy maximum from experimental data, and such a value has been obtained from many of the observed distributions. On the other hand, *in those few cases in which it is possible to predict the energy of the beta decay from data on heavy particle reactions, the visually extrapolated limit has been found to fit the data better than the K-U value.*" They noted, however, the other experimental support for the K-U theory and recorded both the visually extrapolated values as well as those obtained from the K-U theory.

The difficulty of obtaining unambiguous results for the maximum β-decay energy was further illustrated by J. L. Lawson (1939) in his discussion of the history of measurements of the ^{32}P spectrum in which different experimenters obtained quite different experimental results.

The energy spectrum of these electrons was first obtained by J. Ambrosen (1934). Using a Wilson cloud chamber, he obtained a distribution of electrons with an observed upper limit of about 2 MeV (Million electron volts). Alichanow et al. (1936), using tablets of activated ammonium phosphomolybdate in a magnetic spectrometer of low resolving power, find the upper limit to be 1.95 MeV. Kurie, Richardson, and Paxton (1936) have observed this upper limit to be approximately 1.8 MeV. This work was done in a six-inch cloud chamber, and the results were obtained from a distribution involving about 1500 tracks. Paxton (1937) has investigated only the upper regions of the spectrum with the same cloud chamber, and reports that all observed tracks above 1.64 MeV can be accounted for by errors in the method. E. M. Lyman (1937) was the first investigator to determine accurately the spectrum of phosphorus by means of a magnetic spectrometer. The upper limit of the spectrum which he has obtained is 1.7 ± 0.04 MeV.

Lawson's own 1939 value was 1.72 MeV, in good agreement with that of Lyman. The difficulties and uncertainties of the measurements are clear. Measurements made with different techniques disagreed with one another, and physicists may have suspected that the discrepancy was due to the different techniques used. But even measurements made with the same technique differed.

Another developing problem for the K-U theory was that its better fit to the RaE spectrum required a finite mass for the neutrino. This was closely related to the problem of the energy end point because the mass of the neutrino was estimated from the difference between the extrapolated and observed end points. Measurement of the RaE spectrum in the late 1930s had given neutrino masses in the range of 0.3–0.52 m_e, where m_e is the mass of the electron. On the other hand, the upper limit for the neutrino mass from nuclear reactions was less than 0.1 m_e. (The neutrino was thought to have zero mass until very recently. Contemporary experiments show that the neutrino does have mass. The current upper limit for the mass of the neutrino is approximately one-millionth of the mass of the electron.)

Toward the end of the decade, the tide turned and experimental evidence began to favor Fermi's theory over that of Konopinski and Uhlenbeck. Tyler used a thin radioactive source to observe the ^{64}Cu positron spectrum, which reduced the energy lost by the decay electrons in leaving the source. "The thin source results are in much better agreement with the original Fermi theory of beta decay than with the later modification introduced by Konopinski and Uhlenbeck. As the source is made thicker there is a gradual change in the shape of the spectra which gradually brings about better agreement with the K-U theory than with the Fermi theory" (1939).

Similar results were obtained for phosphorus, sodium, and cobalt by Lawson (1939): "In the cases of phosphorus and sodium, where the most accurate work was possible, the shapes of the spectra differ from

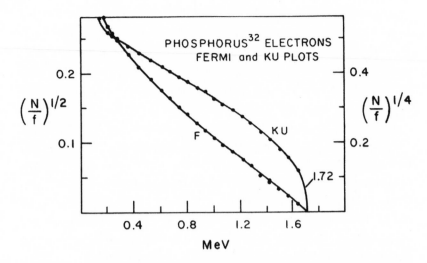

FIGURE 4.3. Fermi and K-U plots for electrons from a thin phosphorus source, ³²P. In this case the Fermi theory fits a straight line and is the better theory. From Lawson (1939).

the results previously reported by other investigators in that there are fewer low energy particles. The reduction in the number of particles has been traced to the relative absence of scattering in the radioactive source and its mounting. The general shape of the spectra is found to agree more satisfactorily with that predicted from the original theory of Fermi than that given by the modification of this theory proposed by Konopinski and Uhlenbeck." The superiority of the Fermi theory is evident. Richardson's earlier warning concerning the dangers of scattering and energy loss in spectrum measurements had been correct. These effects were causing the excess of low-energy electrons. The later, thin-source results for ³²P shown in figure 4.3 are strikingly different from the earlier, thick-source results, also on ³²P, shown in figure 4.2.

Yet another problem was found with the evidential support for the K-U theory. This was pointed out by Lawson and J. Cork in their 1940 study of the spectrum of ¹¹⁴In (indium). Their Kurie plot for the Fermi theory was clearly a straight line indicating that the Fermi theory was correct. They pointed out, "*However, in all of the cases so far accurately presented, experimental results for 'forbidden' spectra have been compared to theories for 'allowed' transitions.* The theory for forbidden transitions [for Fermi's theory] has not been published." (β-decay transitions come in several types, depending on the mathematics that describe the state of the particles involved. Allowed transitions occur more quickly than do forbidden transitions. Fermi had originally calculated the β-decay spectrum only for allowed transitions.) An incorrect experiment-theory comparison had been made. An inappropriate theory had been compared to the experimental results. Similar cautions concerning this type of comparison were made earlier by other experimenters, but little attention was paid to these warnings. At the time the only really good test of the two theories was the β-decay spectrum of ^{114}In, an allowed transition, which allowed a valid comparison between theory and experiment. That valid comparison favored the Fermi theory.

Konopinski and Uhlenbeck Assist Fermi

The spectrum of forbidden transitions for the original Fermi theory was finally calculated by Konopinski and Uhlenbeck in 1941. They noted that some of the evidence from the β-decay spectra that had originally supported their theory now tended to support the Fermi theory: "The authors made a criticism of Fermi's formula on the basis of a comparison with older experimental data and advanced a modification of the Fermi theory which seemed to represent the data better. The technical improvements in the most recent measurements, particularly in eliminating scattering, have withdrawn the basis for the criticism." They remarked that these new measurements had also confirmed the maximum spectrum energy as derived from nuclear masses. "The so-called K-U modification had led to values that were distinctly too large."

They noted, however, that there were still discrepancies between Fermi's theory and other experimental results, so that the choice between the two theories was still unresolved. "Fermi's formula however still does not represent a great number of observed β-spectra. Many of these disagreements are undoubtedly due to the superposition of spectra, as has lately again been emphasized by Bethe, Hoyle, and Peierls.

Nevertheless all the disagreements cannot be explained in this way. The well investigated spectra of RaE and ^{32}P show definite deviations form Fermi's formula," they wrote.

Konopinski and Uhlenbeck attributed the latter discrepancies to the fact the RaE and ^{32}P were forbidden decays. Unlike the case of allowed transitions, for which the shape of the energy spectrum was independent of the mathematical form of the decay interaction, the shape of the forbidden spectra in Fermi's theory did depend on the mathematical form. Konopinski and Uhlenbeck calculated the spectrum shapes for the various possible forms of the interaction for the Fermi theory and noted, "There is, therefore, no a priori reason to expect them to obey the allowed formula." They then compared their calculated spectra to the available experimental results. They reported that they could obtain good fits to the observed spectra of ^{32}P and RaE, thus supporting Fermi's theory

In 1943 Konopinski published a review article on β-decay. He noted, "For β-decay theory, next in importance to the confirmation of the general structure of the theory itself, has been the making of a choice between the Fermi and K-U ansätze. . . . *The K-U criticism and modification of Fermi's theory seems now to be definitely disproved by the following developments."* The evidence cited by Konopinski included the evidence of the β-decay energy spectra. He concluded, "*Thus, the evidence of the spectra, which has previously comprised the sole support for the K-U theory, now definitely fails to support it."*

Only eight years passed from the original publication of the Konopinski-Uhlenbeck theory to the public, and published, declaration by one of its authors that it was incorrect. Konopinski, and the rest of the physics community, agreed that it did not accurately describe β-decay. On the basis of the best experimental evidence available at the time, the observed energy spectra from ^{32}P and RaE decay, Konopinski and Uhlenbeck had proposed a modification of Fermi's theory that better fit that evidence. Experiment had, albeit incorrectly, called for a new theory.

Experimental work continued, and physicists found that these early experimental results were incorrect. It was quickly realized that scattering and energy loss in the radioactive sources used in such experiments had distorted the spectra. Thinner sources were then used, and the new results favored Fermi's theory. At the time, these were technically very difficult experiments. In the early stages of an experimental investigation, it is often difficult to identify sources of background that might mask or mimic the effect one wishes to observe. When physicists realized that scattering and energy loss were a problem, and they did so rather quickly, they took corrective action.

Similarly, the incorrect experiment-theory comparison was eliminated. Konopinski and Uhlenbeck calculated the theoretical spectra needed to solve the allowed-forbidden transition problem. Ironically, after the calculation was done, a correct experiment-theory comparison indicated that Fermi's theory, rather than their own, was correct. In addition, Lawson and Cork performed an experiment on an allowed transition for which a valid experiment-theory comparison could be made. That too favored Fermi's theory. Konopinski and Uhlenbeck, as well as the rest of the physics community, were convinced by the experimental evidence.

(5)

The Rise and Fall of the Fifth Force

The complex history of the fifth force,* a proposed modification of Newton's law of gravity, is an example of the interaction of experiment and theory. The hypothesis was proposed, seemed plausible based on some experimental evidence, and eventually was shown to be incorrect. In this episode the experiments proceeded essentially independently of theory. Once the hypothesis had been proposed and had suggested the size of the effects expected, experimenters wanted to answer the question of whether such a force existed. The experiments had a life of their own. Two different sets of discordant experimental results were produced. In each set, the experimental tests of the fifth-force hypothesis gave conflicting results; one of each set supported the hypothesis, whereas the second found no evidence of such a force.

For a more detailed history, see Franklin (1993b).

*Physicists usually speak of four forces. In decreasing order of strength they are the strong, or nuclear, force, which holds the atomic nucleus together; the electromagnetic force, which holds atoms together; the weak force, responsible for radioactive decay; and the gravitational force.

The Rise

The story of the fifth force begins with a successful experimental test and confirmation of gravitational theory. In 1975, Roberto Colella, Albert Overhauser, and Samuel Werner observed the effect of gravity on a beam of neutrons. Although this experiment showed the presence of gravitational effects at the quantum level, it did not distinguish between general relativity (the accepted theory of gravity) and its competitors. As Ephraim Fischbach (1980) pointed out, this was because the experiment was conducted at low speeds, and at such speeds the predictions of all existing gravitational theories agreed.

Fischbach went on to consider whether gravitational effects might explain the previously observed violation of CP symmetry in K_L^0 decays. CP symmetry is combined parity, or space-reflection, symmetry and charge-conjugation, or particle-antiparticle, symmetry. The K mesons are elementary particles. There are two neutral K mesons, the K_S^0 and K_L^0, the short-lived and long-lived particles, respectively. CP symmetry predicts that the K_S^0 meson, but not the K_L^0 meson, will decay into two π mesons (pions). In 1964 James Cronin, Val Fitch, and their collaborators found that the K_L^0 does decay into two pions (Christenson et al. 1964). This strongly supported the violation of CP symmetry. (Fitch and Cronin later received the Nobel Prize for their work.) Although, as Fischbach noted, there were both experimental and theoretical reasons arguing against gravity as the source of CP violation, the relevance of the arguments to his case were not clear.

Fischbach continued to work on the question of how to observe gravitational effects at the quantum level and how to distinguish between general relativity and its competitors. A high-speed version of the Colella experiment that would use neutrons did not seem feasible, so he turned his attention to K mesons, which appeared to be adaptable to the purpose. He began a collaboration with Sam Aronson, an experimental physicist with considerable experience in K meson experiments. At this time Aronson and his collaborators were investigating the regeneration of K_S^0 mesons. Regeneration was one of the unusual properties of the K^0 mesons. An accelerator-produced beam of K^0 mesons contained 50% K_S^0 mesons and 50% K_L^0 mesons. If all of the K_S^0 mesons were allowed to decay and then the remaining K_L^0 mesons were allowed to interact with matter, the beam once again contained K_S^0 mesons. They had regenerated. The published papers (Roehrig et al. 1977; Bock et al. 1979) stated, "The data are consistent with a constant phase [of the regeneration amplitude]," but Aronson and his colleague Gregory Bock were troubled by what seemed to be an energy dependence of the phase. Although the data were consistent with a constant phase, there was at least a hint of an energy dependence. The low-energy points had a larger phase than the high-energy points.

This suggested energy dependence led Aronson and Fischbach to examine the possible energy dependence of other parameters of the K_S^0-K_L^0 system. They found small but noticeable energy dependencies for all of these quantities. They concluded, "The experimental results quoted in this paper are of limited statistical significance. *The evidence of a positive effect in the energy dependencies of* [the parameters] *is extremely tantalizing, but not conclusive*" (Aronson et al. 1983a).

A second paper by the group sought a theoretical explanation of these effects. They concluded that none of the available explanations was sufficient and stated, "It is clear, however, that if the data . . . are correct, then the source of these effects will represent a new and hitherto unexplored realm of physics" (Aronson et al. 1983b). These results and conclusions were not greeted with enthusiasm or regarded as reliable by everyone within the physics community. Commenting on the need for a new interaction to explain the effects, an anonymous referee of the paper remarked, "This latter statement also applies to spoon bending" (personal communication from Fischbach). The paper was published nonetheless.

A second strand of this story concerns the recent history of alternatives to, or modifications of, standard gravitational theory. Newtonian gravitational theory and its successor, Albert Einstein's general theory of relativity, although strongly supported by existing experimental evidence, have had their competitors. Carl Brans and Robert Dicke (1961), for example, offered an alternative to general relativity. This theory contained a parameter ω that determines the difference between the two theories. For large values of ω, the Brans-Dicke theory is indistinguishable from general relativity, whereas for small values the two theories make quite different predictions. By the end of the 1970s, experimental evidence favored large values of ω and thus favored general relativity.

In the early 1970s Yasunori Fujii suggested a modification of the Brans-Dicke theory, which required a new, and hitherto unobserved, massive particle (1971, 1972, 1974). He found that adding such a particle to the theory gave rise to an additional force that had a short range (10 m–30 km, depending on the details of the model). In Fujii's theory, the gravitational potential had the form $V = -GmM/r[1 + \alpha e^{-r/\lambda}]$, where α was the strength of the new interaction and λ its range. The first term is the ordinary Newtonian gravitational potential, and the second term was Fujii's modification. This model also predicted a gravitational constant G that varied with distance. (Variable constant may seem to be an oxymoron, but it serves as a useful shorthand.) Fujii calculated that in his model the gravitational constant at large distances, G_∞, would be equal to $\frac{3}{4}$ G_{LAB}, the value at short distances.

Fujii also searched for possible experimental tests of his theory. Crucially, he discussed the famous experimental test of Einstein's equivalence principle that had been performed by Roland von Eötvös and his collaborators in the early 20th century and published in 1922. Fujii noted that his new force predicted an effect that was smaller than the upper limit of 1 part in 1 million set by Eötvös, whose experiment was sensitive to such a short-range force. Fujii suggested redoing the Eötvös experiment and also suggested other possible geophysics experiments, although he noted that local mass inhomogeneities might present difficulties. His comments were prescient.

At this time David Mikkelsen and Michael Newman investigated the status of G, the universal gravitational constant (1977). They used data from several sources, including laboratory measurements, and concluded, "Constraints on G(r) in the intermediate distance range from 10 m < r < 1 km are so poor that one cannot rule out the possibility that G_c [G_∞] differs greatly from G_0 [G_{LAB}]." They pointed out that their analysis "does not even rule out Fujii's suggestion that $G_c/G_0 = 0.75$." (Somewhat surprisingly, G is not as well measured as other physical constants.)

The most important summary of work on G, relevant to the subsequent history of the fifth force, was presented by G. W. Gibbons and

B. F. Whiting (1981). Their survey included measurements of gravity in mineshafts and in submarines. The results for G from those measurements were slightly higher than those obtained in the laboratory, but because of experimental uncertainties no firm conclusion could be drawn. Gibbons and Whiting summarized the situation as follows: "It has been argued that our experimental knowledge of gravitational forces between 1 m and 10 km is so poor that it allows a considerable difference between the laboratory measured gravitational constant and its value on astronomical scales—an effect predicted in theories of the type alluded to above [these included Fujii's theory]." Although experiment allowed for such a difference between the laboratory and astronomical values of G, there were reasonably stringent limits on any proposed modification in the distance range 1–10 km. A small window of opportunity existed, however, for a force with a strength approximately 1% that of gravity and with a range between 1 m and 1 km.

Until early 1983 the two strands of this story—the energy dependence of the K meson system parameters and the modifications of Newtonian gravity and their experimental tests—proceeded independently. At about this time Fischbach became aware of the discrepancies between experiment and gravitational theory through the work of Frank Stacey and his collaborators (Stacey and Tuck 1981; Stacey et al. 1981). He made no connection at the time between the two problems because he was still thinking in terms of long-range forces, which had already been ruled out experimentally. In early 1984 he realized that this would not be the case for a short-range force and that the effect could be much smaller. At this time he also became aware of the Gibbons and Whiting summary (1981) and realized that a short-range force might be the common solution to the two problems.

Fischbach, Aronson, and their collaborators looked for other places in which such an effect might be seen. They found only three: (1) the neutral K meson system at high energy, which they had already studied; (2) the comparison of terrestrial and satellite determinations of g, the local gravitational acceleration; and (3) the original Eötvös experiment, which measured the difference between the gravitational and inertial masses of different substances. (There are two types of mass. The inertial mass is a measure of how hard it is to accelerate an object [large masses are harder to accelerate], and the gravitational mass gives rise to the gravitational force.) If a short-range composition-dependent force existed, it might show up in this experiment. (Composition dependence means that the fifth force between two lead masses would be different from the fifth force between a lead mass and a copper mass.) Fischbach, Aronson, and their colleagues reexamined the original data of Eötvös and his colleagues for evidence of a short-range composition-dependent force (Fischbach et al. 1986a). By this time they knew of the recent result that gave G measured in a mine as $G = (6.730 \pm 0.003) \times 10^{-11}$ $m^3kg^{-1}s^{-2}$, in disagreement with the best laboratory value at the time, $G = (6.6726 \pm 0.0005) \times 10^{-11}$ $m^3kg^{-1}s^{-2}$. The difference was still somewhat uncertain, but it was considerably larger than the experimental uncertainties. Fischbach and colleagues used a modified gravitational potential, $V = -GmM/r[1 + \alpha e^{-r/\lambda}]$, and they remarked that such a potential could explain the geophysical data quantitatively if $\alpha = (-7.2 \pm 3.6) \times 10^{-3}$, with $\lambda = 200 \pm 50$ m. This result was within the window that Gibbons and Whiting had found. The potential had the same mathematical form as that suggested much earlier by Fujii, who had also suggested redoing the Eötvös experiment.

Eötvös and his collaborators used a torsion pendulum to measure the difference between the gravitational and inertial masses of various substances.* These masses do not have to be the same. Einstein's theory of general relativity, however, requires that they be identical. The Eötvös experiment is a modern equivalent of Galileo's mythical experiment at the Leaning Tower of Pisa. Galileo was reported to have dropped two unequal masses, with the same composition, and observed

---

*According to one source, the torsion balance was suggested by Juan Hernandez Torsión Herrera, about whom very little is known. He was born of noble parents in Andalusia about 1454. He traveled widely, and on one of his journeys in Granada with his cousin Juan Fernandez Herrera Torsión, both were captured by Moorish bandits. Herrera Torsión died in captivity, but Torsión Herrera managed to escape after a series of magnificent exploits, of which he spoke quite freely in his later years. Although not a scientist in his own right, Torsión Herrera passed on to a Jesuit physicist the conception of his famous Torsión balance. This idea apparently came to him when he observed certain deformations in the machinery involved when another cousin was being broken on the rack. There are several good reasons for doubting the veracity of this story.

that they fell at the same rate. In retrospect, this shows that the inertial and gravitational masses of the substance are equal. The Eötvös experiment tested whether objects made of two different substances fell at the same rate. Eötvös and his collaborators found that the two masses, inertial and gravitational, were equal, to about a part in 1 million, as Einstein's general theory of relativity required (Eötvös, Pekar, and Fekete 1922).

Fischbach and collaborators found that they could explain both the gravitational constant discrepancy and the energy dependence of the K meson parameters with the new force. They also found that Δk, the fractional change in the gravitational acceleration (the difference between the gravitational acceleration of the substance and the average acceleration due to gravity, divided by that average acceleration), would be different for two different substances. They found that Δk was proportional to $\Delta B/\mu$ for the two substances, where B was the baryon number, equal to the number of neutrons and protons of the substance, and μ was the mass of the substance in units of the mass of atomic hydrogen. They reanalyzed Eötvös's original data, plotted the fractional change in   acceleration as a function of $\Delta B/\mu$, and found the surprising results shown in figure 5.1. The data fit a straight line. There was evidence for a composition-dependent force. If there were no such force, then the graph would be a horizontal straight line with Δk = 0. They concluded, "We find that the Eötvös-Pekar-Fekete data are sensitive to the composition of the materials used." They also found that the strength of the new interaction needed to explain the Eötvös data and the geophysical data disagreed by a factor of 15, which they regarded as "surprisingly good" in view of the simple model of the earth they had assumed. Not everyone was so sanguine about this.

The paper by Fischbach and his colleagues can be summarized as follows. A reanalysis of the original Eötvös paper presented suggestive evidence for an intermediate-range composition-dependent force. With a suitable choice of parameters (a force approximately 1% of the gravitational force with a range of approximately 100 m), they could relate this force to measurements of gravity in mineshafts and to a suggested energy dependence in the parameters of the neutral K meson system.

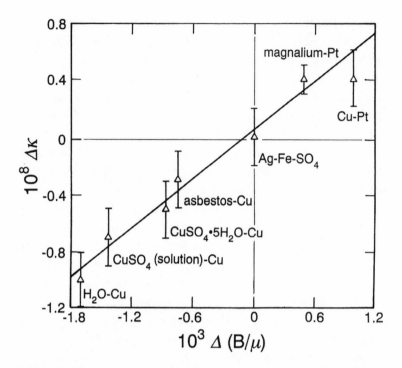

FIGURE 5.1. The fractional change in gravitational acceleration, Δk, as a function of $\Delta B/\mu$, which depends on the substance. A composition dependence is clearly seen. If there were no composition dependence, the graph would be a horizontal straight line with $\Delta k = 0$. From Fischbach et al. (1986a).

The Fall

The suggestion by Fischbach and his colleagues had an immediate impact in the popular press. On January 8, 1986, only two days after the publication of their paper, a headline in the *New York Times* announced, "Hints of Fifth Force in Nature Challenge Galileo's Findings." This referred to the composition dependence of the suggested force, which implied that different substances would fall at different rates and disagreed with what Galileo was supposed to have observed at the Leaning Tower of Pisa. This was the first use of the term "fifth force."

On January 15 an editorial in the *Los Angeles Times* also discussed the subject. It cited the skepticism of Richard Feynman, a Nobel Prize winner in physics. Feynman's skepticism concerned the factor-15 difference (a more careful analysis gave a factor of 30) between the force needed to explain the Eötvös data and that needed to explain the gravitational mine data. Feynman was more bothered by this discrepancy than Fischbach and his colleagues had been. Feynman expressed this concern in a letter published in the January 23 *Los Angeles Times*. Feynman felt that the editorial did not convey his real meaning:

> You reported in an editorial "The Wonder of it All" about a proposal to explain some small irregularities in an old (1909) experiment (by Eötvös) as being due to a new "fifth force." You correctly stated I didn't believe it—but brevity didn't give you a chance to tell why. Lest your readers get to think that science is decided simply by opinion of authorities, let me expand here.
>
> If the effects seen in the old Eötvös experiment were due to the "fifth force" proposed by Prof. Fischbach and colleagues, with a range of 600 feet it would have to be so strong that it would have had effects in experiments already done. For example, measurements of gravity force in deep mines agree with expectations to about 1% (whether this remaining deviation indicates a need for modification of Newton's Law of gravitation is a tantalizing question). But the "fifth force" proposed in the new paper would mean that we have a deviation of at least 15%. This calculation is made in the paper by the authors themselves (a more careful analysis gives 30%). Although the authors are aware of this (as confirmed in a telephone conversation) they call this "surprisingly good agreement," while it, in fact, shows they cannot be right.
>
> Such new ideas are always fascinating, because physicists wish to find out how Nature works. Any experiment which deviates from expectations according to known laws commands immediate attention because we may find something new.

Feynman agreed that there were possible ways to make the results agree, but he regarded them as unlikely. Fischbach answered by pointing out the important effects of local mass asymmetry, which were not known when Feynman wrote his letter.

The battle would not, however, be conducted or decided either in the popular press or in private correspondence, but rather in the technical literature. During 1986 considerable attention was devoted to the status of the fifth-force hypothesis. Questions were raised about whether the reanalysis of the Eötvös data was valid and whether the proposed new force was already ruled out by experiment. Evidence from other previous experiments was also cited, and possible theoretical explanations and implications of the new force were examined. New experiments, as well as improvements in the sensitivity of existing experiments, were suggested.

One of the most important developments was the recognition that local mass asymmetries, such as cliffs, hills, or large buildings, were of crucial importance not only in the reanalysis of the Eötvös experiment, but also in the design of experiments to search for the fifth force. This was first noted by Fischbach and his collaborators. They remarked that further analysis had shown "that one cannot in fact deduce from the EPF [Eötvös-Pekar-Fekete] data whether the force is attractive or repulsive. The reason for this is that in the presence of an intermediate-range force, local *horizontal* mass inhomogeneities (e.g., buildings or mountains) can be the dominant source in the Eötvös experiment" (Fischbach et al. 1986b). To determine the magnitude and sign of the effect, more detailed knowledge of the local mass distribution was needed than was then available. Fischbach and his collaborators even searched for a detailed map of the University of Budapest campus, where Eötvös had done his work. They also tried to discover whether the building in which the experiment was done had a basement, which would influence the local mass distribution. The importance of the local mass distribution could also explain the numerical discrepancy between the force derived from the Eötvös reanalysis and that found from the mine data that had bothered Feynman and others. Other authors suggested redoing the Eötvös experiment by placing the torsion balance on a high cliff or in a tunnel in such a cliff. They claimed that such a location, with its large local mass asymmetry (on the edge of a cliff there is mass on one side of the experimental apparatus, but not on the other), could increase the sensitivity of the experiment by a factor of 500. Alvaro De Rujula quipped (1986), "Although malicious rumor has it that Eötvös himself weighed more than 300 pounds [suggesting that Eötvös himself was the source of a local mass asymmetry], unspecific hypotheses are not, *a priori,* particularly appealing." (Eötvös was a mountain climber, and photographs indicate that he did not weigh 300 lb.)

The initial reanalysis of the Eötvös experiment was partially incorrect because it did not consider local mass asymmetries. The subsequent criticism not only modified the theoretical model but also allowed the design of experiments that would be far more sensitive to the presence of the hypothesized fifth force. Other critics suggested that there was, in fact, no observed effect and that Fischbach and his colleagues had made an error in the reanalysis. De Rujula, however, performed his own reanalysis of the Eötvös data and obtained results identical to those of Fischbach and collaborators. Some physicists suggested that experiments on K mesons had already ruled out the fifth force. Questions were also raised about whether the Eötvös data could be explained in terms of more conventional physics, without invoking a new force.

Although the criticism may have made the reanalysis of the Eötvös data somewhat uncertain, it did not prevent physicists from planning new, more sensitive versions of old experiments and designing new ones to test for the presence of the fifth force. At the same time, theoretical physicists were attempting to find an explanation for the force and to see if it had implications in other areas. Unfortunately, in all of these theoretical studies the expected effects were quite small and did not suggest new experimental tests.

At the end of 1986, the evidential context for the fifth force was much the same as it had been at the beginning of the year, on January 6, when Fischbach and colleagues had first proposed it. By early 1986 the inverse square law of gravity had been tested at very short distances and had been confirmed, but the possibility of an intermediate-range force remained. Doubts had been raised about the proposed mechanism of the force, but other explanations were possible. The tantalizing effects of the reanalysis of the Eötvös experiment, the K meson parameters, and the measurements of gravity in mineshafts remained.

The attitude of scientists toward the fifth force at this time varied from rejecting it outright to finding it highly suggestive and plausible. Sheldon Glashow, a Nobel laureate and theoretical physicist, was quite negative: "Unconvincing and unconfirmed kaon data, a reanalysis of the Eötvös experiment depending on the contents of the Baron's wine cellar [an allusion to the importance of local mass inhomogeneities], and a two-standard-deviation geophysical anomaly! Fischbach and his friends offer a silk purse made out of three sows' ears, and I'll not buy it" (quoted in Schwarzschild 1986). John Maddox noted (1986), "Fischbach et al. have provided an incentive for the design of better measurements by showing what kind of irregularity it will be sensible to look for." An important feature of experimental design is knowing how large the observed effect is supposed to be. A much more positive view was also heard: "Considerable, and justified, excitement has been provoked by the recent announcement—that a reanalysis of the Eötvös experiment together with recent geophysical gravitational measurements supports the existence of a new fundamental interaction" (Lusignoli and Pugliese 1986).

It seems clear, judging by the substantial amount of work published in 1986, that a significant segment of the physics community thought the fifth-force hypothesis was plausible enough to be worth further investigation. Although almost invisible in the published literature, experiments were being designed, performed, and analyzed. The results began to appear in early 1987.

Was Galileo Wrong?

Two sets of discordant experimental results had to be resolved to decide whether there was a fifth force. The first strand of experimental investigation of the fifth force was the search for a composition dependence of the gravitational force. These were, in fact, the first published experimental results. The strongest piece of evidence cited when the fifth force was originally proposed came from a reanalysis of the Eötvös experiment. That reanalysis had shown a large and surprising composition-dependent effect. This was the effect that was subsequently investigated.

Two types of composition-dependence experiments are shown in figure 5.2. To observe the effect of a short-range force such as the fifth force, a local mass asymmetry is needed. This asymmetry was provided by either a terrestrial source—a hillside or a cliff—or by a large, local, laboratory mass. A composition-dependent, short-range force would cause the torsion pendulum made of two different substances to twist. A variant was the float experiment, in which an object floated in a fluid and the difference in gravitational force on the float and on the fluid

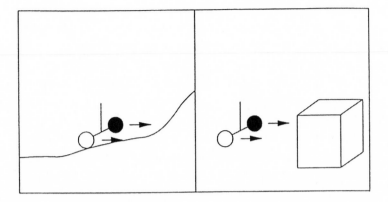

FIGURE 5.2. Two types of composition-dependence experiments. *Left*, a terrestrial source; *right*, a laboratory source. From Stubbs (1990).

would be detected by the motion of the float. Experiments of this type were done with terrestrial sources.

The results of the first tests for a composition-dependent force appeared in January 1987, 1 year after the fifth force was first proposed. They disagreed. Peter Thieberger, using a float experiment, found results consistent with the presence of such a force. A group at the University of Washington, headed by Eric Adelberger and whimsically named the Eöt-Wash group, found no evidence for such a force and set rather stringent limits on its presence (Adelberger et al. 1987).

The results of Thieberger's experiment, performed on the Palisades cliff in New Jersey, are shown in figure 5.3. The float moves quite consistently and steadily away from the cliff (the y-direction), as expected if there was a fifth force. (One wag remarked that all the experiment showed was that any sensible float wanted to leave New Jersey.) Thieberger eliminated other possible causes for the observed motions. He also rotated his apparatus by 90° to check for possible instrumental asymmetries and obtained the same positive result. In addition, he performed the same experiment at another location, one without a local mass asymmetry or cliff, and found no effect, as expected. He concluded, "The present results are compatible with the existence of a medium-range, substance-dependent force which is more repulsive (or less attractive) for Cu than for H_2O.... Much work remains before the

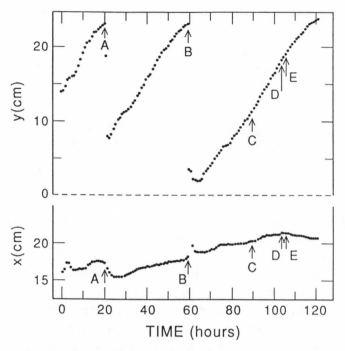

FIGURE 5.3. The position of the center of Thieberger's float as a function of time. There is a definite motion away from the cliff (the y axis). From Thieberger (1987).

existence of a new substance-dependent force is conclusively demonstrated and its properties fully characterized."

The Eöt-Wash experiment used a torsion pendulum located on the side of a hill on the University of Washington campus. If the hill attracted the copper and beryllium test bodies that were used in the apparatus differently, then the torsion balance would experience a net torque. None was observed (figure 5.4). The group also eliminated other possible causes of effects that might either mimic the presence of a fifth force or mask the effects of such a force.

The discordant results were an obvious problem for physicists. Both experiments appeared to be carefully done, with all plausible and significant sources of possible error and background adequately accounted for. Yet the two experiments disagreed.

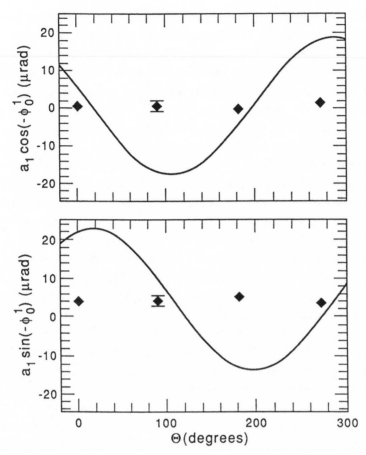

FIGURE 5.4. Deflection signal for the Eöt-Wash torsion balance. The curves are the predictions of the fifth force. Experiment and theory clearly disagree. From Raab (1987).

In this case, attempts were made to observe and measure the same quantity, a composition-dependent force, with very different apparatuses, a float experiment and a torsion pendulum. Scientists wondered whether some unknown but crucial background in one of the experiments produced the wrong result. To this day, no one has found an error in Thieberger's experiment, but the consensus is that the Eöt-Wash group is correct and that Thieberger is wrong—that there is no fifth force.

The discord was resolved by an overwhelming preponderance of evidence. The torsion pendulum experiments were repeated by others, including Val Fitch (Fitch, Isaila, and Plamer 1988), Ramanath Cowsik (Cowsik et al. 1988, 1990), William Bennett (1989), Paul Boynton (1990), and Riley Newman (Newman, Graham, and Nelson 1989; Nelson, Graham, and Newman 1990), and by the Eöt-Wash group (Stubbs et al. 1989; Heckel et al. 1989). None found evidence for a fifth force.

Bennett's experiment is particularly interesting. He reported a measurement of the difference in force exerted on copper and lead masses by a known mass of water, located nearby. The experiment used a torsion balance located near the Little Goose Lock on the Snake River in eastern Washington, in which the water level was changed periodically to allow the passage of boats. This change in water level provided the known mass of water. The difficulty of real, as opposed to ideal, experiments is clearly illustrated in this experiment. "Because the data were taken during a dry period (August 1988), separate lock fillings could not be made just for the experiment. On average there were four 'lockages' a day from barge traffic which could occur at any hour of the day or night with only a half-hour advance notice." The apparatus needed minor adjustment every 4 or 5 hours and then took about 2 hours to stabilize, allowing good data to be taken for the next 2 or 3 hours. "The success of a particular run depended on the coincidence of this observation period with the arrival of 'locktraffic' and typically only one could be observed in a period of about 6 h during weekdays. Fortunately, traffic on weekends was heavier because of pleasure craft. Although consistent with individual isolated experiments, by far the best data were obtained on Sunday, 21 August 1988, when an armada of small craft went up and down the river." Bennett found no evidence of the fifth force.

All the repetitions, in different locations and with different substances, gave consistently negative results. Evidence against the fifth force also came from modern versions of Galileo's Leaning Tower of Pisa experiment, performed by Kazuaki Kuroda and Norikatsu Mio (1989a, 1989b) and by James Faller and his collaborators (Niebauer, McHugh, and Faller 1987). As more negative evidence was provided, the initial, and startling, effect claimed by Fischbach and collaborators be-

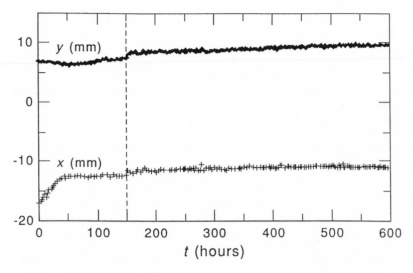

FIGURE 5.5. Position of Bizzeti's float as a function of time. No effect is seen. From Bizzeti et al. (1988).

came less and less dramatic. In fact, the effect disappeared. In addition, Bizzeti, using a float apparatus similar to the one used by Thieberger, obtained results showing no evidence of a fifth force (Bizzeti et al. 1989). Bizzeti's results (figure 5.5) can be compared with those of Thieberger (figure 5.3). Bizzeti's result was quite important. Had he agreed with Thieberger, then the question would arise of whether some systematic difference between torsion-balance experiments and float experiments gave rise to the conflicting results. This did not happen. An overwhelming preponderance of evidence was against composition dependence of the fifth force. Even Thieberger, although he had not found any error in his own experiment, agreed. "Unanticipated spurious effects can easily appear when a new method is used for the first time to detect a weak signal. . . . Even though the sites and the substances vary, effects of the magnitude expected have not been observed. . . . It now seems likely that some other spurious effect may have caused the motion observed at the Palisades cliff" (1989). He was persuaded that he was wrong by the evidence.

Towers and Mineshafts

A second way to test for the presence of the fifth force was by investigating the distance dependence of the gravitational force, to see if there was a deviation from Newton's inverse-square law. This type of experiment measured the variation of gravity with position, usually in a tower, sometimes in a mineshaft or borehole. All the experiments used a standard device, a LaCoste-Romberg gravimeter, to measure gravity. The measurements were then compared with the values calculated with a model of the earth, surface gravity measurements, and Newton's law of gravitation. This type of calculation had been done often and was regarded as reliable. The results of the calculation were, however, quite sensitive to the surface gravity measurements and to the model of the earth used. This made knowledge of the local mass distribution and of the local terrain very important.

Evidence from such measurements had provided some of the initial support for the existence of the fifth force. Geophysical measurements during the 1970s and 1980s had given values of G, the universal gravitational constant, that were consistently higher, by approximately 1%, than that obtained in the laboratory. Because of possible local mass anomalies, they were also tantalizingly uncertain.

New experiments followed the proposal of the fifth force. At the Moriond workshop in January 1988, Donald Eckhardt presented results from the first of the new tower gravity experiments. The results differed from the predictions of the inverse-square law by -500 ± 35 µGal (1 µGal $= 10^{-8}$ ms^{-2}) at the top of the tower (figure 5.6).

Further evidence for the fifth force was provided by a group that measured the variations in gravity in a borehole in the Greenland icecap (Ander et al. 1989). They found an unexplained difference between the measurements taken at a depth of 213 m and those taken at a depth of 1,673 m. The experimental advantage of the Greenland experiment was the uniform density of the icecap. The disadvantages were the paucity of surface gravity measurements and the presence of underground geological features that could produce gravitational anomalies.

All the evidence from tower and mineshaft experiments before 1988 supported the fifth force. There was, however, considerable although

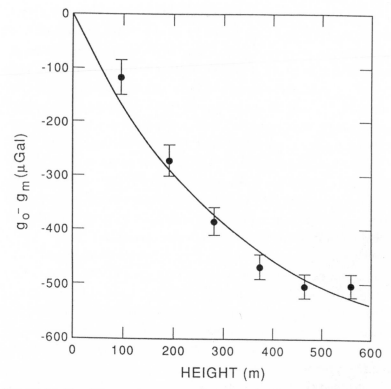

FIGURE 5.6. The difference between measured and calculated values of the acceleration due to gravity as a function of height obtained by Eckhardt et al. (1988). There is a clear discrepancy between the experimental results and the predictions of Newtonian gravitational theory. From Fairbank (1988).

not unambiguous negative evidence from other types of experiment. Negative evidence from tower experiments would also be forthcoming.

Even before those negative results appeared, questions and doubts were raised concerning the positive results. It was not the gravity measurements themselves that were questioned. These were all obtained with a standard and reliable instrument. Rather, the theoretical calculations used for the theory-experiment comparison were criticized. One of the important features needed in these calculations was an adequate model of the earth.

The Greenland researchers' calculation was the first to be criticized, and severely, particularly for the paucity of surface gravity measurements near the location of their experiment (their survey included only

16 such points) and for the inadequacy of their model of the earth. It was pointed out that underground features existed in Greenland of the type that could produce such gravitational anomalies. The group later admitted that their result could either be interpreted as evidence for non-Newtonian gravity (a fifth force) or explained by local density variations. "We cannot unambiguously attribute it to a breakdown of Newtonian gravity because we have shown that it might be due to unexpected geological features below the ice" (Ander et al. 1989).

Robert Parker, a member of the Greenland group, as well as David Bartlett and Wesley Tew (1989a, 1989b), suggested that both the positive evidence for the fifth force of Eckhardt and collaborators and that from the mineshaft experiments could be explained by either local density variations or inadequate modeling of the local terrain.

Eckhardt disagreed. He and his group presented a revised, and lower, value for the deviation from Newtonian gravity at the top of their tower of 350 ± 110 μGal. They attributed this change, a reduction of approximately one-third, to better surface gravity data and to finding a systematic elevation bias in their previous survey. (Gravity measurements tend to be made on roads, rather than in ditches or surrounding fields. Roads are usually higher than their surroundings, causing an elevation bias.) "We also had the help of critics who found our claims outrageous." They concluded, "Nevertheless the experiment and its reanalysis are incomplete and we are not prepared to offer a final result" (Eckhardt et al. 1989).

A group from the Lawrence Livermore Laboratory presented a result from their gravity measurements at the BREN tower at the Nevada test site. To overcome the difficulties with their previous calculations, they had extended their gravity survey to include 91 of their own gravity measurements within 2.5 km of the tower, supplemented with 60,000 surface gravity measurements within 300 km that were done by others. This is far more than the 16 points in the Greenland survey. They presented preliminary results in agreement with Newtonian gravity, reporting that, at the top of the tower, there was no difference between the measured and predicted values (Kasameyer et al. 1989; Thomas et al. 1989).

Bartlett and Tew continued their work on the effects of local terrain.

They argued that the Hilton mine results of Stacey and his collaborators could also be caused by a failure to include local terrain in their theoretical model. They communicated their concerns to Stacey privately (1989b). Their view was confirmed when, at the General Relativity and Gravitation Conference in July 1989, G. J. Tuck reported that his group had incorporated a new and more extensive surface gravity survey into its calculation. "Preliminary analysis of these data indicates a regional bias that reduces the anomalous gravity gradient to two thirds of the value that we had previously reported (with a 50% uncertainty)." With such a large uncertainty, the results of Stacey and his collaborators could no longer be considered as support for the fifth force.

Parker and Mark Zumberge, two members of the Greenland group, offered a general criticism of tower experiments. They argued, in some detail, that they could explain the anomalies reported in both Eckhardt's tower experiment and their own ice cap experiment with conventional physics and plausible local density variations. They concluded that there was "no compelling evidence for non-Newtonian long-range forces in the three most widely cited geophysical experiments [those of Eckhardt and Stacey and their own] . . . and that the case for the failure of Newton's Law could not be established" (1989).

The last hurrah for tower gravity experiments that supported the fifth force was signaled in the paper "Tower Gravity Experiment: No Evidence for Non-Newtonian Gravity" (Jekeli, Eckhardt, and Romaides 1990). In this paper Eckhardt and his group presented their final analysis of their data, which included a revised theoretical model, and concluded that there was, in fact, no deviation from Newtonian gravity. (Their results, in figure 5.7, contrast with their initial positive result shown in figure 5.6). Two subsequent tower results also supported Newton's law (Speake et al. 1990; Kammeraad et al. 1990).

The discord had been resolved. The tower and mineshaft measurements were correct. The comparison between theory and experiment had led to the discord. It had been shown that the results supporting the fifth force could be explained by inadequate theoretical models, either failure to account adequately for local terrain or failure to include plausible local density variations.

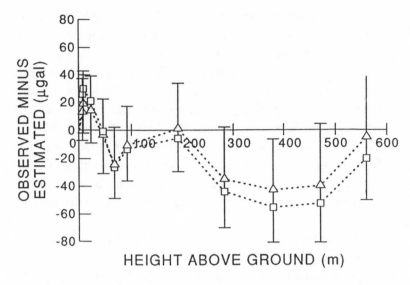

FIGURE 5.7. The difference between measured and calculated values of the acceleration due to gravity as a function of height found by Eckhardt's group in 1990. The fifth force effect has disappeared. From Jekeli, Eckhardt, and Romaides (1990).

Scientists make decisions in an evidential context. The fifth force was a modification of Newtonian gravity. Newtonian gravity, and its successor, general relativity, were strongly supported by other existing evidence. In addition, other credible negative tower gravity results did not suffer from the same difficulties as did the positive results. There was also an overwhelming preponderance of evidence against the fifth force from other types of experiment. The decision as to which theory-experiment comparison was correct was not made solely on the basis of the experiments and calculations themselves, although this would have been justified. Scientists examined all of the available evidence and came to a reasoned decision about which were the correct results—and concluded that the fifth force did not exist.

In 1990, at a Moriond Workshop attended by most of those working in the field, Orrin Fackler of the Livermore group remarked, "The Fifth Force is dead." No one disagreed. There is no fifth force.

(a)

II

THE SEARCH FOR

WHAT IS THERE

(6)

The Discovery of the Electron

Early belief in the existence of the electron was prompted by experiments conducted in the late 19th and early 20th centuries. Joseph John (J. J.) Thomson's 1897 experiment on cathode rays (figure 6.1) is generally regarded as the discovery of the electron.

The Cathode Ray Experiment

Cathode rays were discovered in the late 19th century. They were emitted from the electrically negative plate (cathode) of an evacuated tube and traveled to the positive plate (anode). Thomson set out to investigate the nature of this new phenomenon. He was attempting to decide between two views: the rays were negatively charged material particles or they were disturbances in the ether, the hypothetical fluid in which electromagnetic waves propagated.

For more details, see Smith (1997), which includes some of Thomson's original papers, and Franklin (1997a).

FIGURE 6.1. J. J. Thomson (1856–1940), photographed some time around the turn of the 20th century. Courtesy American Institute of Physics Emilio Segre Photo Archive.

The experiments discussed in this paper were undertaken in the hope of gaining some information as to the nature of Cathode Rays. The most diverse opinions are held as to these rays; according to the almost unanimous opinion of German physicists they are due to some process in the aether to which—inasmuch as in a uniform magnetic field their course is circular and not rectilinear—no phenomenon hitherto observed is analogous: another view of these rays is that, so far from being wholly aetherial, they are in fact wholly material, and that they mark the paths of particles of matter charged with negative electricity. (Thomson 1897)

Thomson's first order of business was to show that the cathode rays carried negative charge. That had presumably been shown earlier by Jean Perrin. Perrin had placed two coaxial metal cylinders, insulated from one another, in front of a plane cathode. Each cylinder had a small hole through which the cathode rays could pass. The outer cylinder was grounded. When cathode rays passed into the inner cylinder, an electroscope attached to it showed the presence of a negative electrical charge. When the cathode rays were magnetically deflected so that they did not pass through the holes, no charge was detected. "Now the supporters of the aetherial theory," Thomson wrote, "do not deny that electrified particles are shot off from the cathode; they deny, however, that these charged particles have any more to do with the cathode rays than a rifle-ball has with the flash when a rifle is fired."

Thomson repeated the experiment in 1897, but in a form that was not open to that objection. His apparatus is shown in figure 6.2. Like Perrin's, it had two coaxial cylinders with holes. The outer cylinder was grounded and the inner one attached to an electrometer, to detect any charge. The rays passed from the cathode (A) into the larger bulb, but they did not enter the holes in the cylinders unless they were deflected by a magnetic field. Thomson concluded,

> When the cathode rays (whose path was traced by the phosphorescence on the glass) did not fall on the slit, the electrical charge sent to the electrometer when the induction coil producing the rays was set in action was small and irregular; when, however, the rays were bent by a magnet so as to fall on the slit there was a large charge of negative electricity sent to the electrometer. . . . If the rays were so much bent by the magnet that they overshot the slits in the cylinder, the charge passing into the cylinder fell again to a very small fraction of its value when the aim was true. *Thus this experiment shows that however we twist and deflect the cathode rays by magnetic forces, the negative electrification follows the same path as the rays, and that this negative electrification is indissolubly connected with the cathode rays.* (Emphasis added)

There was, however, a problem for the view that cathode rays were negatively charged particles. Several experiments, in particular one by Heinrich Hertz, had failed to observe the deflection of cathode rays by an electrostatic field. If the cathode rays were electrically charged, then they should have been deflected by such a field. Thomson proceeded to answer this objection with the apparatus shown in figure 6.3. Cathode rays from the cathode in the small bulb at the left passed through a slit in the anode and then through a second slit. They then passed between the two plates and produced a narrow, well-defined phosphorescent patch at the right end of the tube, which also had a scale attached to measure any deflection.

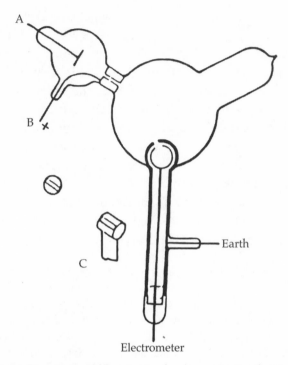

A

B

C

Earth

Electrometer

FIGURE 6.2. J. J. Thomson's 1897 apparatus for demonstrating that cathode rays have negative electric charge. The slits in the cylinders are shown. Adapted from Thomson (1897).

When Hertz had performed the experiment, he had found no deflection when a potential difference (electrostatic field) was applied across the two plates. He therefore concluded that the electrostatic properties of the cathode rays are either nil or very feeble. Thomson admitted that when he first performed the experiment, he, too, saw no effect. But "on repeating this experiment [that of Hertz] I at first got the same result [no deflection], but subsequent experiments showed that the absence of deflexion is due to the conductivity conferred on the rarefied gas by the cathode rays. On measuring this conductivity it was found that it diminished very rapidly as the exhaustion increased; it seemed that on trying Hertz's experiment at very high exhaustion there might be a chance of detecting the deflexion of the cathode rays by an electrostatic force" (Thomson 1897). (Thomson continued his investigation of the conductivity of gases and won the Nobel Prize for

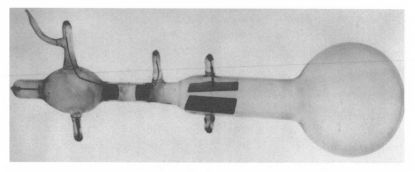

FIGURE 6.3. Thomson's tube for demonstrating that cathode rays are deflected by an electric field. It was also used to measure e/m. Courtesy of Cavendish Laboratory.

that work, not for the discovery of the electron.) Thomson then performed the experiment at lower pressure (higher exhaustion) and observed the deflection. He also demonstrated that the cathode rays were deflected by a magnetic field. Thomson concluded, "As the cathode rays carry a charge of negative electricity, are deflected by an electrostatic force as if they were negatively electrified, and are acted on by a magnetic force in just the way in which this force would act on a negatively electrified body moving along the path of these rays, I can see no escape from the conclusion that they are charges of negative electricity carried by particles of matter." Thomson was using the well-known duck argument. If it looks like a duck, quacks like a duck, and waddles like a duck, there is good reason to believe that it is a duck.

Having established that cathode rays were negatively charged material particles, Thomson went on to ponder the nature of the particles. "What are these particles? Are they atoms, or molecules, or matter in a still finer state of subdivision?" To investigate this question, Thomson made measurements of the charge-to-mass ratio of cathode rays.

Thomson's method used both electrostatic and magnetic deflection of the cathode rays. His apparatus was essentially the same as the one he used to demonstrate the electrostatic deflection of cathode rays (figure 6.3). He could apply a magnetic field perpendicular to both the electric field and the trajectory of the cathode rays. By adjusting the strengths of the electric and magnetic fields so that the cathode ray beam was undeflected, Thomson determined the velocity of the rays.

Turning off the magnetic field allowed the rays to be deflected by the electric field. From the measured deflection, the length of the apparatus, and the electric and magnetic field strengths, Thomson could calculate the ratio m/e for cathode rays.

He found a mass-to-charge ratio m/e of $(1.29 \pm 0.17) \times 10^{-8}$ gm/C (grams per coulomb). (The modern value is 0.56857×10^{-8} gm/C.) This ratio appeared to be independent of both the gas in the tube and the metal used in the cathode, suggesting that the particles were constituents of the atoms of all substances. It was also far smaller, by a factor of 1,000, than the smallest mass-to-charge ratio previously measured for the hydrogen ion in electrolysis.

Thomson remarked that this surprising result might be due to the smallness of m or to the bigness of e. He argued that m was small, citing Philipp Lenard, who had shown that the range of cathode rays in air (half a centimeter) was far larger than the mean free path of molecules in air (approximately 10^{-5} cm). If the cathode ray traveled so much farther than a molecule before colliding with an air molecule, it must be very much smaller than a molecule. If it was smaller, then it should have a smaller mass. Thomson concluded that the negatively charged particles were also constituents of atoms.

In sum, Thomson had shown that cathode rays behaved as negatively charged material particles would be expected to behave. They deposited negative charge on an electrometer and were deflected by both electric and magnetic fields in the appropriate direction for a negative charge. In addition, the value for the mass-to-charge ratio was far smaller than the smallest value previously obtained, for the hydrogen ion. If the charge was the same as that on the hydrogen ion, the mass would be far less. In addition, the cathode rays traveled farther in air than did molecules, also implying that they were smaller than an atom or molecule. Thomson concluded that these negatively charged particles were constituents of atoms.

Millikan and the Oil Drop

Thomson did not use the term "electron" to refer to his negatively charged particles; he preferred "corpuscle." "Electron" was introduced by the Irish physicist G. Johnstone Stoney in 1881 as the name of the "natural unit of electricity," the amount of electricity that must pass through a solution to liberate one atom of hydrogen. Stoney did not associate the electron with a material particle, and physicists at the time questioned whether electricity might not be a continuous homogeneous fluid. Lord Kelvin, for example, raised this question and commented, "I leave it however, for the present and prefer to consider an atomic theory of electricity . . . largely accepted by present day workers and teachers. Indeed Faraday's laws of electrolysis seem to necessitate something atomic in electricity" (1897).

The early determinations of the charge of the electron had not established whether there was a fundamental unit of electricity. That was because the experiments measured the total charge of a cloud of droplets, without showing that the value obtained was anything other than a statistical average. The same was true for Thomson's measurement of e/m for a beam of cathode rays.

It was the experimental work of Robert Millikan (figure 6.4) at the University of Chicago, beginning in 1909, that provided the next step in establishing the electron as a fundamental particle. Millikan's great improvement in experimental technique was to perform all of the measurements on a single charge carrier. "In a preceding paper a method of measuring the electrical charge was presented which differed essentially from methods which had been used earlier by other observers only in that all of the measurements were made upon one individual charged carrier. This modification eliminated the chief sources of uncertainty which inhered in preceding determinations by similar methods such as those made by Sir Joseph Thomson, H. A. Wilson, Ehrenhaft, and Broglie, all of whom had deduced the elementary charge from the average behavior in electrical and gravitational fields of swarms of charged particles" (Millikan 1911). Millikan not only established that there was a fundamental unit of electrical charge, but also measured it accurately. He associated that fundamental unit with the charge both on Thomson's corpuscles and on the hydrogen ion in electrolysis. Combining this with the measurements of e/m for both objects determined that the mass of Thomson's corpuscle was $1/1845$ that of the hydrogen atom. This was yet another step in establishing the electron as a fundamental particle. It had both a definite mass and a

FIGURE 6.4. Robert Millikan (1868–1953) as a postdoctoral student in Germany in 1895–1896. Courtesy of California Institute of Technology.

definite charge, and it behaved exactly as a negatively charged particle would be expected to behave.

Millikan's experimental apparatus is shown in figure 6.5. He allowed single oil drops to fall a known distance in air and measured the time of fall. (The drops were falling at a constant terminal velocity.) He then turned on an electric field and measured the time it took for each drop to travel the same distance upward under the influence of the electric field. These two time measurements let him determine both the mass of the drop and its total charge.

The charge on the oil drop sometimes changed spontaneously by absorption of charge from the air or by ionization. Millikan also induced such changes with either a radioactive source or x-radiation. The change in charge on a drop from successive times of ascent with the field

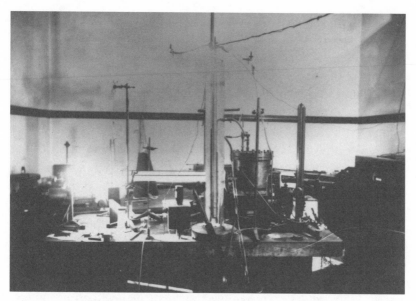

FIGURE 6.5. Millikan's oil-drop apparatus for measuring the charge of the electron. Courtesy of California Institute of Technology.

on could also be calculated. Millikan found that both the total charge of the drop and the changes in that charge were small integral multiples of e, a fundamental unit of charge.

Millikan wrote, "The total number of changes which we have observed would be between one and two thousand, and *in not one single instance has there been any change which did not represent the advent upon the drop of one definite invariable quantity of electricity or a very small multiple of that quantity*" (1911). He concluded that he had succeeded in his goal "to present direct and tangible demonstration, through the study of the behavior in electrical and gravitational fields of this oil drop, carrying its captured ions, of the correctness of the view advanced many years ago and supported by evidence from many sources that all electrical charges, however produced, are exact multiples of one definite, elementary charge." Millikan's final value for e was $(4.774 \pm 0.009) \times 10^{-10}$ esu (electrostatic units; Millikan 1913). (The modern value is $4.80320420 \times 10^{-10}$ esu.)

Despite his claim to the contrary, Millikan did not publish all of his oil-drop results. He excluded many drops because he was not sure that the apparatus was working properly, some because of experimental or calculational difficulties, some because they simply weren't needed (he had far more data than he needed to improve the measurement of e by a factor of 10), and a few apparently for the sole reason that they increased the experimental uncertainty. One drop, which gave a value of e that was 40% low, was also excluded. For that one, Millikan wrote "won't work" in his laboratory notebook. I speculate that this exclusion was simply to avoid giving Felix Ehrenhaft ammunition in the then-current controversy concerning whether there was a fundamental unit of electric charge. Later analysis has shown that the data for this drop were indeed unreliable.

Millikan also discarded some of the data from accepted drops, and he engaged in selective calculation. But the effects of all this cosmetic surgery were quite small. If all the good data are included and all the calculations are done as advertised, the value of e changes by only a part in a thousand, with insignificant increase in the experimental uncertainty.*

Millikan associated his measured e both with the charge on Thomson's corpuscles and the charge on the hydrogen ion in electrolysis. He combined his value for e with contemporary measurements of e/m by electrolytic and cathode ray techniques to determine that the mass of Thomson's corpuscle was $1/1845$ that of the hydrogen atom—surprisingly close to $1/1837.15$, the modern value. Now both a definite mass and a definite charge had been determined for this would-be fundamental particle, and it behaved as a negatively charged particle would be expected to behave.

Scientists now had good evidence for believing that it was a constituent of atoms—in other words the electron.

The Bohr Theory

In 1913, not long after Millikan's oil-drop results were published, Niels Bohr was constructing a theory of the atom, whose confirmation would provide support for the view that the electron was both a fundamen-

*For more details on Millikan's selectivity, see chapter 11.

tal particle and a constituent of atoms. Bohr began with Rutherford's nuclear model of the atom, with a small, massive, positively charged nucleus orbited by electrons of charge e and mass m. Noting that classical electrodynamics would not allow such a system to be stable, he postulated that the electron could, nonetheless, exist in stationary orbits without radiating energy. He calculated the binding energy of the states of an electron orbiting the atomic nucleus.

He further assumed that the electron emitted radiation only when it made a transition from one stationary state to another and that the transition energy when the electron moved from the nth to the n'th state was in the form of a light quantum of energy $E = h\nu = W_n - W_{n'}$, where ν is the frequency of the quantum of light, h is Planck's constant, and W_n and $W_{n'}$ are the energies of the nth and n'th states.

This resulted in a formula that not only fitted the observed Balmer series in hydrogen, the spectrum of light emitted, but also harmonized with the empirical constant used to fit that spectrum. Using the best available values for the electric charge and other physical constants, Bohr calculated that the spectroscopic proportionality constant in the Balmer formula, N, was equal to 3.1×10^{15} s^{-1}, in good agreement with 3.290×10^{15} s^{-1}, the measured spectroscopic value at the time (Bohr 1913).

Somewhat later, Millikan discussed the same issue (1917, 209–10).

The evidence for the soundness of the conception of non-radiating electronic orbits [Bohr's theory] is to be looked for, then, first, in the success of the constants involved. . . . If these constants come out right within the limits of experimental error, then the theory of non-radiating electronic orbits has been given the most crucial imaginable of tests, especially if these constants are accurately determinable.

What are the facts? The constant N of the Balmer series in hydrogen . . . is known with the great precision attained in all wave-length determinations and is equal to 3.290×10^{15}. From the Bohr theory [one can also calculate this value]. As already indicated, I recently redetermined e with an estimated accuracy of one part in 1,000 and obtained for it the value 4.774×10^{-10}. . . . [Using the best available values for physical constants one calculates] $N = 3.294 \times 10^{15}$, *which agrees within a tenth of 1 per cent with the observed value.* This agreement constitutes the most extraordinary justification of the theory of non-radiating electronic orbits.

Millikan could barely contain his enthusiasm for Bohr's theory. He challenged critics to present an alternative that fit the experimental results: "It demonstrates that the behavior of the negative electron in the hydrogen atom is at least correctly described by the *equation* of a circular non-radiating orbit. If this equation can be obtained from some other physical condition than that of an actual orbit, it is obviously incumbent on those who so hold to show what that condition is. Until this is done, it is justifiable to suppose that the equation of an orbit means an actual orbit." Obviously Millikan did not expect them to be able to provide another answer.

Both Millikan and Bohr thought that the existence of the electron, as both a fundamental particle and as a constituent of atoms, was already so well established that they did not even argue that this spectacularly successful prediction supported it. Instead, they argued that the result supported the more controversial assumptions of Bohr's theory, such as the existence of stationary states. This is a good example of the view that to have good reason for holding a theory is, ipso facto, to have good reason for believing in the existence of the entities postulated by that theory.

That argument might be appear to have been discarded when the Bohr theory was superseded a decade later by the quantum mechanics of Schrödinger and Heisenberg, but in fact it was not. The Schrödinger equation also assumes an electron with charge e and mass m and gives exactly the same prediction as the Bohr theory for the Balmer series.

The Stern-Gerlach Experiment and the Discovery of Electron Spin

In the early 1920s, another intrinsic property of the electron, that of spin, emerged. (One can think of electron spin, also called intrinsic angular momentum, as resembling the earth spinning on its axis. The orbital angular momentum is like the earth revolving about the sun.) In 1921, Otto Stern and Walter Gerlach used the already known properties of the electron to design an experiment to search for spatial quantization of electrons in atomic states, as predicted by Arnold Sommerfeld's elaboration of the Bohr theory (1916a, 1916b). This is the idea that the orbital angular momentum of the electrons in such a state can point only in certain specified directions. (Angular momentum is a vector quantity, which has both a magnitude and a direction.) Gerlach and Stern wrote (1921), "The experiment, if it can be carried out, [will result] in a clear-cut decision between the quantum-theoretical and the classical view."

Sommerfeld's theory, along with the measured properties of the electron, acted as an enabling theory for the experiment. It provided an estimate of the size of the magnetic moment of the atoms so that Stern could begin calculations to see if the experiment was feasible. Stern calculated, for example, that a magnetic field gradient of 10^4 gauss (a unit of magnetic field strength) per centimeter would be sufficient to produce deflections that would give detectable separations of the beam components. He asked Gerlach if he could produce such a gradient. Gerlach said yes, and he could do even better. In the experiment, silver atoms passed through the inhomogeneous magnetic field. If the beam was spatially quantized, as Sommerfeld predicted, two spots should be observed on the screen. (According to Sommerfeld, the angular momentum would point in one of two directions: along the magnetic field or opposite to the magnetic field.)

A preliminary result reported by Stern and Gerlach (1921) did not show splitting of the beam into components. It did, however, show a broadened beam spot. "The three experiments with the magnetic field yielded a spot, broadened in the direction of $\partial H/\partial z$ [the magnetic field gradient], having a height of 0.1mm (millimeter) and a width of 0.25 to 0.30mm. One still cannot recognize with certainty an intensity structure within this band. The amount of the symmetric broadening corresponds to a magnetic moment of 1 to 2 Bohr magnetons. . . . On the basis of our experience thus far, we do not doubt that we shall be in a position to decide about the directional quantization through experiments with beams of smaller diameter and perhaps a better method of developing." They concluded that although they had not demonstrated spatial quantization, they had provided "evidence that the silver atom possesses a magnetic moment." (A charged particle with angular momentum behaves as if it were a small magnet.)

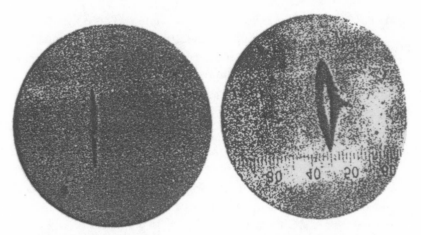

FIGURE 6.6. The results of the Stern-Gerlach experiment. *Right*, with the magnetic field on. There is an intensity minimum in the center of the pattern, and the separation of the beam into two components is clearly seen. This indicates that the electron has spin $1/2$. *Left*, with the magnetic field off. No deflection of the beam is seen. From Gerlach and Stern (1922).

Stern and Gerlach made improvements in the apparatus. The results are shown in figure 6.6. There is an intensity minimum in the center of the pattern, and the separation of the beam into two components is clearly seen. "Apart from any theory, it can be stated, as a pure result of the experiment, and as far as the exactitude of our experiments allows us to say so, that silver atoms in a magnetic field have only *two discrete* values of the component of the magnetic moment in the direction of the field strength; both have the same absolute value with each half of the atoms having a positive and a negative sign respectively" (Gerlach and Stern 1922).

This result confirmed Sommerfeld's quantum-theoretical prediction of spatial quantization. The notoriously skeptical Wolfgang Pauli wrote to Gerlach (February 17, 1922), "Hopefully now even the incredulous Stern will be convinced about directional quantization." (Stern had thought that their preliminary result had argued against spatial quantization.) Pauli's view was generally shared by the physics community.

Stern and Gerlach concluded that they had established the existence of spatial quantization. A few years later, following the suggestion of electron spin by Samuel Goudsmit and George Uhlenbeck (Uhlenbeck and Goudsmit 1925, 1926), it was realized that the Stern-Gerlach experiment actually provided evidence for an electron with spin ½.

It might reasonably be claimed that electron spin was discovered before it was invented.

Is It the Same Electron?

In the 1920s the charge e of the electron was $(4.774 \pm 0.009) \times 10^{-10}$ esu. Its mass m was $1/1845$ that of the hydrogen atom and it had a magnetic moment indistinguishable from one Bohr magneton. A recent edition of *Review of Particle Physics*, a 974-page blockbuster, reveals that the charge of the electron is $(4.80320420 \pm 0.00000019) \times 10^{-10}$ esu (Hagiwara et al. 2002). The electron mass is $(9.10938188 \pm 0.00000072) \times 10^{-31}$ kg, approximately $1/1837.15$ of the mass of the hydrogen atom. The magnetic moment is $(1.0011596521869 \pm 0.0000000000041)$ μ_B, where μ_B is the Bohr magneton, in exquisite agreement with what quantum electrodynamics predicts, 1.001159652460.

Allowing for improvements in both the precision and accuracy of these measurements, the properties of the electron have remained constant. Much has been learned about the electron and its properties and interactions in the intervening time, but its defining properties have stayed the same. It is still a negatively charged particle with a definite charge and a definite mass. It has spin ½ and is a constituent of atoms. The electron, as an entity, has remained constant even though the theories we use to describe it have evolved dramatically.

Are There Really Electrons?

Bas van Fraassen is a philosopher of science who does not believe that we can have good reasons for belief in the existence of particles such as the electron (1980, 74–77). It is possible to see van Fraassen, hear him, and in other ways detect his presence with unaided human senses; he can also be described (hair and eye color) and measured (height and

FIGURE 6.7. The author on his way to visit an electron, which in addition to being the lightest charged particle, is also one of the smallest towns in western Washington State.

weight). That would surely be convincing proof that there is such a real person. It would be bizarre, then, to say only, "The world is such that everything is as if there were a real Bas van Fraassen."

The electron, on the other hand, is an entity that can be observed only with instruments. Yet why should such special status be given to unaided human sense perception? Sense perception can, on occasion, be quite unreliable: for example, there are the effects of mirages, drugs, dreams, and sleep deprivation. Eyewitness identifications in jury trials are very unreliable. "Is it live or is it Memorex?" asked a television ad for a brand of audiotape.

Most people believe that "seeing is believing," and that no argument is needed for the correctness of human sense perception. I believe they are wrong. The same arguments, the epistemology of experiment, that should be made to validate a sense perception are precisely the same as those that should be, and are, provided to show the validity of instrumental observation. If the existence of a real Bas van Fraassen is to be believed, then the same credence should be accorded to electrons. Figure 6.7 shows the author on his way to visit an electron.

(7)

The Road to the Neutrino

In his 1960 poem "Cosmic Gall," John Updike describes neutrinos:

> Neutrinos, they are very small.
> They have no charge and have no mass
> And do not interact at all.

Updike's description of the properties of neutrinos is reasonably accurate. The best experimental evidence available at the time showed that neutrinos had no electrical charge and no mass. It was also known that in addition to being electrically neutral, the neutrino has an intrinsic angular momentum, or spin. If the neutrino is pictured as a small sphere, then its intrinsic angular momentum is like the earth spinning on its axis. Updike's only misstatement is that neutrinos do not interact at all with matter. The neutrino does interact with matter, but it does so very weakly. In fact, its interaction length in lead should be measured in light-years. Today, even more is known about the

For a more detailed history of the neutrino, including developments up to the present, see Franklin (2000), which includes arguments for the existence of the neutrino and how physicists learned so much about this elusive particle.

properties of the neutrino. The neutrino is also left-handed (it has a handedness like a screw), and there are three kinds of neutrinos. There is also good evidence that one kind of neutrino can transform into another kind of neutrino and that the neutrino may have a very small mass, about one-millionth of the mass of the electron. The neutrino can even be used as a tool to investigate other aspects of nature.

The usual history of the neutrino hypothesis found in physics textbooks goes something like this:

Radioactivity—the spontaneous transformation of one element into another—produces α-particles or β-particles or γ-rays. Experimental work on the energy of the electrons emitted in β-decay began early in the 20th century, and the observations posed a problem: If only two bodies (the daughter nucleus and an electron) are present in the final state of β-decay, the conservation of energy and momentum require that the energy spectrum of decay electrons must be monoenergetic. Thus, the observation of a continuous spectrum—electrons emitted with all energies from zero up to a maximum energy that depended on the radioactive element—cast doubt on both of these conservation laws. Or perhaps the electrons lost varying amounts of energy in escaping the radioactive substance, thus accounting for the continuous energy spectrum. But careful experiments showed that this was not the case. So the problem persisted. In the early 1930s, Wolfgang Pauli suggested that an undetected neutral particle was also emitted in β-decay. Enrico Fermi (1934a, 1934b) dubbed this particle the neutrino. That solved the problem of the continuous energy spectrum, because in a three-body decay the energy of the electron is no longer required to be unique. The energy and momentum conservation laws were saved.

This version of the story has several virtues. It is clear, and it seems to have an almost inevitable logic. Physicists proceeded from the observation of a continuous energy spectrum in β-decay, by means of the application of the conservation laws of energy and momentum, to the need for a new, low-mass, neutral particle that was also emitted in β-decay. The only problem with this story is that it is both incorrect and incomplete.

What really happened is far more complex and far more interesting. The first difficulty with the story as usually told concerns the observation of the continuous energy spectrum in β-decay. The process by which physicists eventually came to the conclusion that the spectrum was continuous took some 30 years. It was not a simple matter of measuring the energy of the electrons emitted in β-decay.

The Discovery of Radioactivity

Radioactivity was almost accidentally discovered by Henri Becquerel in 1896. His work was stimulated by Wilhelm Röntgen's discovery of x-rays a few months earlier. Becquerel had been working on phosphorescence, the delayed emission of light by a substance after it has been exposed to an external source of light. After Röntgen's announcement, Becquerel began an investigation of whether phosphorescent substances would emit x-rays if they were exposed to intense light. His initial experiments produced no effects, but when he used uranium salts, which he had prepared for phosphorescence experiments 15 years earlier, he found a striking effect (figure 7.1). Becquerel described his experiment as follows: "One wraps a photographic plate . . . in two sheets of very thick black paper . . . so that the plate does not fog during a day's exposure to sunlight. A plate of the phosphorescent substance is laid above the paper on the outside and the whole is exposed to the sun for several hours. When the photographic plate is subsequently developed, one observes the silhouette of the phosphorescent substance, appearing in black on the negative [figure 7.1]. If a coin, or a sheet of metal . . . is placed between the phosphorescent material and the paper, then the image of these objects can be seen to appear on the negative" (1896b).

The plate in the figure shows a dark smudge, not very convincing evidence of anything. For Becquerel, however, it was a stimulus for further investigation. He concluded, "The phosphorescent substance in question emits radiations which penetrate paper that is opaque to light" (1896b). One week later, Becquerel admitted that his earlier interpretation of his result was wrong. He published a paper demonstrating that his observed phenomenon had nothing to do with phosphorescence (1896c). William Crookes, a British physicist who often worked with Becquerel, described the discovery (1909, xxii):

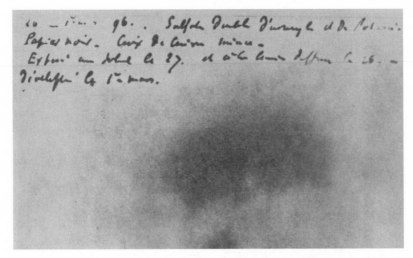

FIGURE 7.1. Becquerel's first evidence for radioactivity was this photographic plate, which was wrapped in opaque black paper and placed under a piece of uranium salt on February 26, 1896. From Segre 1980.

Becquerel was working on the phosphorescence of uranium compounds after insolation [exposure to sunlight]; starting with the discovery that sun-excited uranium nitrate gave out rays capable of penetrating opaque paper [his earlier result] and then acting photographically, he had devised another experiment in which, between the plate and the uranium salt, he interposed a sheet of black paper and a small cross of thin copper. On bringing the apparatus into daylight the sun had gone in, so it was put back into the dark cupboard and there left for another opportunity of insolation. But the sun persistently kept behind clouds for several days, and, tired of waiting (or with the unconscious prevision of genius), Becquerel developed the plate. To his astonishment, instead of a blank, as expected, the plate had darkened under the uranium as strongly as if the uranium had been previously exposed to sunlight, the image of the copper cross shining out white against the black background.

Becquerel observed the same effect with several uranium salts, from which he inferred that the effect was due to the presence of uranium. He confirmed this by an experiment in which he used only pure uranium metal and obtained the same result. He concluded that uranium was emitting a form of radiation that could both penetrate opaque paper and expose a photographic plate. Subsequent experiments, by the Curies and others, showed that other substances, including the newly discovered elements radium and polonium, emitted similar radiation. What that radiation actually was, however, remained an unanswered question.

The first step in deciphering the nature of the radiation emitted by uranium was taken by Ernest Rutherford in 1899. He measured the intensity of the radiation emitted by uranium as a function of the thickness of aluminum foils placed over the uranium. He found that, at first, each plate reduced the amount of radiation by the same, constant fraction, but that beyond a certain thickness the intensity of the radiation was only slightly reduced by adding additional layers. Rutherford concluded, "These experiments show that the uranium radiation is complex, and that there are present at least two distinct types of radiation —one that is very readily absorbed, which will be termed for convenience the α radiation, and the other of a more penetrative character, which will be termed the β radiation."

It was soon shown that the α-particles were positively charged, the β's negatively charged, and the γ-rays—a third type of emitted radiation discovered by Paul Villard in 1900—electrically neutral. Subsequent work by Rutherford and others showed that the α-particles were helium nuclei. The γ-rays were found to be high-energy electromagnetic radiation.

In 1904 William Bragg argued that α-particles of the same initial energy or velocity had the same range in matter, an important point for later work. Bragg measured the ionization produced by a pencil-like beam of α-particles as they passed through air.

> In the case when all the rays are initially of uniform velocity, the curve obtained ought to show, when the radium is out of range of the ionization chamber, an effect due entirely to β and γ rays, which should slowly increase as the distance diminishes. When the α rays can just penetrate, there should a somewhat sudden appearance be of the ionization, and for a short distance of the approach, equal to the depth of the chamber, the curve should be a parabola. Afterwards it should become a straight line.

This is exactly realized; and so far the hypothesis is verified. But a further effect appears. As the radium is gradually brought nearer to the chamber, the straight line suddenly changes its direction; and indeed there appear to be two or three such changes. . . .

For all this there is a ready explanation. The atom passes through several changes, and it is supposed that at four of these an α atom is expelled. Probably the α particles due to one change are all projected with the same speed.

Bragg had shown not only that α-particles were emitted with the same energy in each particular radioactive decay, but also that they had a constant range in matter.

What then of the β-rays? As early as 1900 experiments by Becquerel, by Friedrich Giesel, by Stefan Meyer and Egon von Schweidler, and by Pierre and Marie Curie had found that β-rays had the same negative charge as that of cathode rays. At about the same time (1901–1906), Walter Kaufmann began a series of experiments on β-rays emitted from a radium source. In 1902 he concluded, *"For small velocities, the computed value of the mass of the electrons which generate Becquerel rays . . . fits within observational errors with the value found in cathode rays"* (emphasis added). Other experiments at the time confirmed Kaufmann's result. From then on, the physics community identified β-rays with cathode rays. They were both electrons.

The caveat about small velocities in Kaufmann's claim concerning the electron mass was important. He had found that radium emits electrons with a wide range of velocities, up to almost the speed of light. He had used those high-speed electrons to investigate the variation of electron mass with velocity. This was a significant question at the time (see Miller 1981 for details). In the early 20th century, several theoretical physicists, including Max Abraham and Alfred Bucherer, had attempted to explain the origin of the mass of the electron and had derived an expression for the variation of the electron's mass with its velocity. Hendrick Lorentz and Albert Einstein, using the principle of relativity on which Einstein's special theory of relativity was based, had also calculated such an effect. The three expressions differed. Kaufmann's results seemed to favor the theories of Abraham and Bucherer and to disagree with that of Lorentz and Einstein. Kaufmann's results were, in fact, so credible that Lorentz wrote, in a 1906 letter to Henri Poincare, "Unfortunately my hypothesis of the flattening of electrons is in contradiction with Kaufmann's results, and I must abandon it. I am, therefore, at the end of my Latin" (quoted in Miller 1981, 334).

Einstein agreed but was more sanguine. "The theories of the electron's motion of Abraham and Bucherer [agree better with Kaufmann's data] than the relativity theory. In my opinion both theories have a rather small probability" (1907). Other physicists urged caution and suggested that Kaufmann's analysis of his data might be incorrect and that there might be unknown sources of uncertainty in his experiment. This turned out to be correct. Later experiments, particularly by Adolf Bestelmeyer and by Bucherer, not only supported the Einstein-Lorentz theory, but also pointed to difficulties in Kaufmann's experiment. The evidence supported the special theory of relativity. Still, the question remained, what was the energy spectrum of electrons emitted in β-decay?

The Energy Spectrum in β-Decay

Kaufmann's experiments had demonstrated that radium emitted electrons with a wide range of velocities. A similar result was also found by Becquerel in 1900. Despite the evidence provided by both Kaufmann and Becquerel, the physics community did not accept that the energy spectrum of electrons emitted in β-decay was continuous. There were, at the time, plausible reasons for this widespread skepticism. Physicists argued that the sources used by both Kaufmann and Becquerel were not pure β-ray sources, but contained several elements, each of which could emit electrons with different energies. In addition, even if the electrons started out monoenergetic, they might well lose different amounts of energy in escaping from the radioactive source.

This view was due, in part, to a faulty analogy with α-decay. All of the α-particles from the decay of a particular nucleus do have the same, unique energy, as well as a definite range in matter. At that time it was generally thought that the β-decay spectrum would also be monoenergetic. It was known that β-particles of the same energy, unlike α's, did not have a unique range in matter. It was expected that such β-rays would obey an exponential absorption law. (If particles obey an exponential absorption law and if one-half of an initial beam of particles is absorbed by a certain amount of matter, then the addition of an equal amount of material will reduce the beam to one-quarter of its initial intensity.) As William Bragg stated (1904), "Nevertheless it is clear that β-rays are liable to deflexion through close encounters with the electrons of atoms; and therefore the distance to which any given electron is likely to penetrate before it encounters a serious deflexion is a matter of chance. This, of course, brings in an exponential law."

Experimental work on electron absorption in the first decade of the 20th century, particularly by Heinrich Schmidt (1906, 1907) and by Lise Meitner and Otto Hahn (Hahn and Meitner 1908a, 1908b, 1909a, 1909b) gave support to such a law and therefore to the homogeneous nature of β-rays. Schmidt (1906, 1907) claimed to be able to fit the electron absorption curves for electrons emitted from different radioactive substances with either a single exponential or with a superposition of a few exponentials. Figure 7.2 shows the absorption curve that Schmidt obtained for electrons from radium B (^{214}Pb) and from radium C (^{214}Bi). (If the logarithm of the intensity plotted as a function of the amount of absorbing material is a straight line, then the absorption is exponential.) It was this association of monoenergetic electrons with an exponential absorption law that informed early work on the β-decay spectrum. There was, of course, a circularity to that reasoning. Physicists believed that, if electrons were monoenergetic, their absorption would be exponential. Conversely, therefore, if they were absorbed exponentially, it followed that they were monoenergetic. As Ernest Rutherford remarked, "Since Lenard had shown that cathode rays . . . are absorbed according to an exponential law, it was natural at first to assume that the exponential law was an indication that the β-rays were *homogeneous*, i.e., consisted of β-particles projected with the same speed. On this view, the β-particles emitted from uranium which gave a nearly exponential law of absorption, were supposed to be homogeneous. On the other hand, the β-rays from radium which did not give an exponential law of absorption were known from other evidence to be heterogeneous" (1913, 209–10).

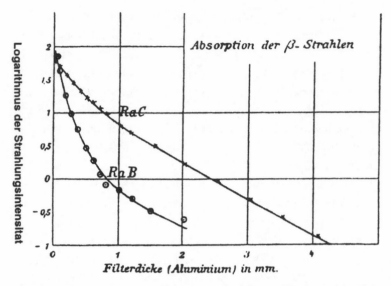

FIGURE 7.2. Schmidt's result on the absorption of β-rays. The logarithm of the electron intensity (ionization) as a function of the absorber can be fit to two straight lines, indicating an exponential absorption law. This suggested that there were two groups of electrons emitted in the decay, each with a different energy. From Schmidt (1906).

The situation changed dramatically with the work of William Wilson (1909). He was investigating what was, in retrospect, a glaring omission in the existing experimental program: the actual measurement of the velocity dependence of electron absorption. He noted that his "present work was undertaken with a view to establishing, *if possible*, the connection between the absorption and velocity of β rays. *So far no actual experiments have been performed on this subject*" (emphasis added). Wilson was right.

He selected a beam of β-rays, all with the same momentum and energy, and measured their absorption. He found that the absorption was linear rather than exponential (figure 7.3). He wrote, "Without entering at present into further details, it can be stated that the ionisation [the electron intensity] did not vary exponentially with the thickness of matter traversed. But, except for a small portion at the end of the curve, followed approximately a linear law." This result contradicted those of Schmidt and of Meitner, Hahn, and Otto von Baeyer. Wilson was also able to demonstrate that the previous experiments, which appeared to follow an exponential law, had been misinterpreted. The others were, in fact, observing the absorption of electrons with a continuous energy spectrum. Exponential absorption, Wilson calculated, required a continuous energy spectrum. Wilson later demonstrated this conclusion experimentally by passing a monoenergetic beam of electrons through an absorber, to make the spectrum continuous, and then measuring their absorption. It was indeed exponential.

Wilson had not, however, demonstrated that the energy spectrum of β-decay electrons was continuous. It would not have been difficult. All he needed to do was measure the ionization produced in his experiment as a function of the electron velocity with no absorber present. He didn't do it, perhaps because he was concerned primarily with the problem of absorption.

Meitner, Hahn, and von Baeyer, as well as the rest of the physics community, agreed that Wilson was correct. They improved their experimental apparatus and began to examine the β-decay energy spectrum (Hahn and Meitner 1910; von Baeyer, Hahn, and Meitner 1911). Electrons emitted from a radioactive source were bent in a magnetic field, passed through a small slot, and were then recorded on a photographic plate. Electrons of the same energy would follow the same path and produce a single line on the photographic plate. The resulting spectrum, shown in figure 7.4, still seemed to support their view that there was one unique electron energy for each of the two radioactive elements contained in the source. The experimenters wrote, "The present investigation shows that, in the decay of radioactive substances, not only α-rays but also β-rays leave the radioactive atom with a velocity characteristic for the species in question. This lends new support to the hypothesis of Hahn and Meitner" (von Baeyer and Hahn 1910). Subsequent work, however, showed a large number of such lines in the β-decay spectra of single elements, thus rendering the one element/one energy hypothesis that Meitner, Hahn, and von Baeyer had proposed untenable. Rutherford, for example, reported 29 lines in the spectra of radium B + radium C (Rutherford and Robinson 1913). As Hahn later wrote, "Our earlier opinions were beyond salvage. It was impossible to assume a separate substance for each beta line" (1966, 57).

In fact, Rutherford and his collaborators showed that the line spectrum of β-rays was really just a secondary effect produced by γ-rays.

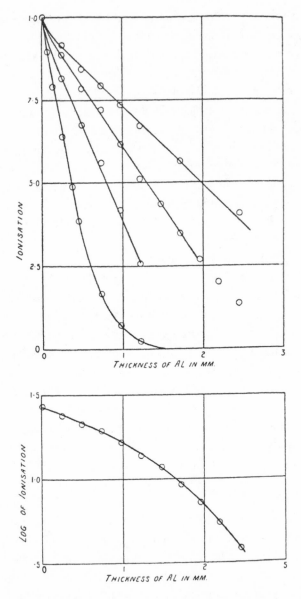

FIGURE 7.3. The upper graph shows the ionization, not its logarithm, as a function of absorber. It is a straight line, indicating a linear, not an exponential, absorption law. The lower graph shows the logarithm of the ionization as a function of absorber. If the absorption were exponential, this graph would be a straight line. From Wilson (1909).

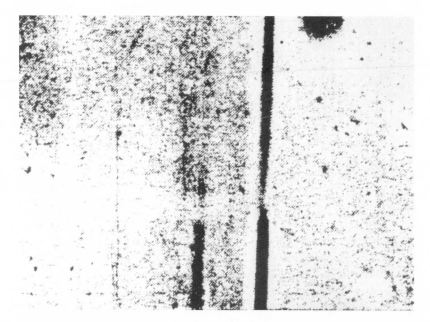

FIGURE 7.4. The first line spectrum for β-decay published by Meitner, Hahn, and von Baeyer. The two observed lines were thought to be produced by the two radioactive elements present in the source. From von Baeyer and Hahn (1910).

The primary β-decay electron energy spectrum was still unknown—but not for long.

In a letter to Rutherford, James Chadwick, who had worked with Rutherford at Manchester University and then had gone on to work with Hans Geiger in Berlin, hinted at the solution. "We [Geiger and Chadwick] wanted to count the β-particles in the various spectrum lines of RaB + C and then to do the scattering of the strongest swift groups. I get photographs very quickly easily, but with the counter I can't even find the ghost of a line. There is probably a silly mistake somewhere" (1914).

Using Geiger's newly invented counters, they could not reproduce the line spectra they and others had seen on photographic plates. This was not a failure of a new experimental apparatus, but rather a problem with the earlier measurements. Chadwick soon reported that the β-decay energy spectrum was continuous (figure 7.5). Chadwick (1914)

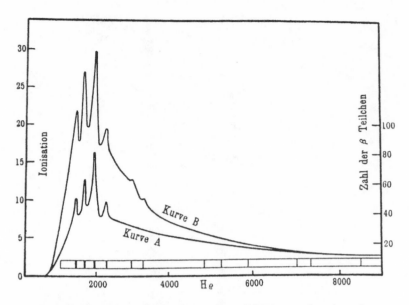

FIGURE 7.5. Chadwick's results for the number of β-rays as a function of energy. A few discrete lines are seen above a continuous energy spectrum. Curve A was obtained with a Geiger counter. Curve B was obtained with an ionization chamber. From Chadwick (1914).

saw four lines, identical to some found in previous spectral measurements, superimposed on a larger continuous energy spectrum.

Why Did It Take So Long?

That the continuous spectrum had been missed by all of the earlier experiments was, in large part, due to an artifact of the photographic detection method. Rutherford had commented earlier that the photographic method could enhance the presence of weak electron energy lines against the continuous background caused by γ-rays and scattered electrons. The photographic method seemed to enhance the perception of the discrete lines in contrast to the background continuum.

Despite the apparent decisiveness of Chadwick's experiment, not everyone within the physics community accepted the observed continuous energy distribution as the primary spectrum of β-decay electrons. In part, that was because no other experimenter had replicated Chadwick's result, with either a radium source, such as the one Chadwick had used, or another radioactive element. And perhaps more decisive was the absence of any theoretical explanation for a continuous spectrum.

Following a break in scientific activity caused by World War I, both experimental and theoretical work on the problem resumed. Chadwick, an Englishman, was detained in Germany during the war. During his internment, he became acquainted with Charles Ellis, another English internee, and enlisted his help in performing scientific experiments. Ellis so enjoyed this work that he gave up his ambition of a career as an artillery officer to pursue a career in science. He would go on to perform extremely important experiments on β-decay.

Meitner (1922a, 1922b) argued against the continuous spectrum on both experimental and theoretical grounds. She noted the complex nature of the β-decay spectrum. In her view it contained many lines, some of which were made diffuse by the scattering of decay electrons from atoms. She argued that Chadwick saw a continuous energy spectrum simply because his experimental apparatus lacked sufficient energy resolution to resolve the lines. Meitner, citing her own previous work with Hahn and von Baeyer, also argued that a quantized system such as an atomic nucleus was unlikely to emit a continuous spectrum. The quantum mechanics that had recently been proposed required that the atomic nucleus or an electron in an atom occupy only certain discrete energy states. Energy is released only when the atom undergoes a transition from one such state to another. The energy difference is also discrete and can take on only certain values. This accounts for the discrete line spectra of the light emitted by atoms. If the nucleus that emitted the electron in β-decay was in one quantum state and the resulting nucleus was in another quantum state, then physicists believed that the electron emitted should also have a discrete, and unique, energy. This was indirectly supported by the fact that the γ-rays emitted in radioactive decay have such a discrete spectrum, similar to that of the light emitted by atoms.

Ellis and William Wooster disagreed with Meitner. They argued that her proposed mechanisms for β-ray energy loss—Compton scattering, bremsstrahlung, and scattering from atomic electrons—could not quantitatively account for that energy loss. (Compton scattering is the scattering of γ-rays by electrons. Bremsstrahlung is the emission of electromagnetic radiation by an electron.) Having eliminated Meitner's plausible alternative explanations of the phenomenon, Ellis and Wooster concluded (1925), "We are left with the conclusion that the disintegration electron is actually emitted from the nucleus with a varying velocity. We are not able to advance any hypothesis to account for this but we think it important to examine what this fact implies." They also noted that there was, in fact, a direct test of whether the primary electrons lost energy as they escaped from either the atom or the entire source: "This is to find the heating effect of the β-rays from radium E (^{210}Bi). If the energy of every disintegration is the same then the heating effect should be between 0.8 and 1.0 $\times$ 10^6 V per atom and the problem of the continuous spectrum becomes the problem of finding the missing energy. It is at least equally likely that the heating effect will be nearer 0.3 $\times$ 10^6 V per atom, that is, will be just the mean kinetic energy of the disintegration electrons." They wrote that they were engaged in performing this experiment, but they suggested that it would be some time before they had definitive results.

One possible explanation that Ellis and Wooster rejected was the possibility that energy was not conserved exactly in each β-decay, but only statistically conserved in an ensemble of many decays. "The next point is to consider how this inhomogeneity of velocity has been introduced. We assume that energy is conserved exactly in each disintegration, since if we were to consider the energy to be conserved only statistically there would no longer be any difficulty in the continuous spectrum. *But an explanation of this type would only be justified when everything else had failed, and although it may be kept in mind as an ultimate possibility, we think it best to disregard it entirely at present*" (emphasis added). Others did not regard this possibility as so far-fetched. There was, of course, another possible explanation, one that Ellis and Wooster did not consider: that another, neutral particle was emitted in β-decay.

In 1927 Ellis and Wooster presented the definitive experimental result they had promised earlier. They firmly established that the energy spectrum of electrons emitted in β-decay was continuous. They did this by measuring the heating effect produced by β-decay electrons in radium E, to determine the average disintegration energy of these elec-trons. If the energy spectrum really was continuous, then the average energy obtained from the heating measurements would equal the average energy obtained by other methods. If, on the other hand, the primary spectrum was monoenergetic and the observed spectrum was due to unknown energy losses, then the average measured heating energy would be at least as large as the maximum energy of the observed continuous energy spectrum.

For radium E the average and maximum energies were 390,000 eV (electron volts) and 1,050,000 eV, respectively. The average heating energy found was 344,000 ± 40,000 eV, in good agreement with the average value of 390,000 ± 60,000 eV obtained by ionization measurements, and in marked disagreement with the value of more than 1 million V expected for monoenergetic electrons. Ellis and Wooster concluded, "We may safely generalise this result obtained for radium E to all β-ray bodies, and the long controversy about the origin of the continuous spectrum of β-rays appears to be settled."

In 1930, Meitner and Wilhelm Orthmann repeated the heating-effect experiment with an improved apparatus and obtained an average energy per β-particle of 337,000 ± 20,000 eV, in excellent agreement with Ellis and Wooster. Meitner wrote to Ellis, "We have verified your results completely. It seems to me now that there can be absolutely no doubt that you were completely correct in assuming that beta radiations are primarily inhomogeneous. But I do not understand this result at all" (quoted in Sime 1996, 105).

Meitner wasn't the only one. The energy spectrum of electrons emitted in β-decay was continuous. It had taken 30 years from the discovery of radioactivity by Becquerel to establish this fact. The question concerning the continuous β-decay energy spectrum had been answered, but the difficulties were just beginning. No one knew why there was a continuous spectrum. If the decay was to a two-body state, as physicists at the time believed, then the conservation of energy and of momentum required a unique energy for the electron emitted—which was clearly not the case. The conservation laws were under attack.

There were two major responses to the establishment of the continuous energy spectrum in β-decay. The experimental result of Ellis and Wooster had revived speculation about the nonconservation of energy. In 1924 Niels Bohr and others had incorporated nonconservation of energy into a quantum theory of radiation (Bohr, Kramers, and Slater 1924). Experiments on the Compton effect had argued strongly against this view. George Thomson (1928) and Davisson and Germer (1927) had demonstrated the wave characteristics of the electron, an entity that is usually considered a particle. Thomson felt that he had to resort to extraordinary measures to explain this: "Some of these questions I should like very briefly to discuss, but we now leave the sure foothold of experiment for the dangerous but fascinating paths traced by the mathematicians among the quicksands of metaphysics." He explicitly cited the result of Ellis and Wooster and wrote, "We are thus reduced to suppose either that the conservation of energy does not apply to each individual process, or that among the atoms . . . there are some individuals with a million volts more energy than others, or that there is some way at present unknown by which the atoms can equalize their energies."

Bohr was also stimulated by the result of Ellis and Wooster to resurrect his earlier views on the nonconservation of energy. Although these were not made public until his Faraday Lecture in 1930 (published in 1932), he had, in 1929, expressed similar views in letters to other physicists and in his private manuscripts. Bohr noted the risk he was taking: "The loss of the unerring guidance which the conservation principles have hitherto offered in the development of atomic theory would of course be a very disquieting prospect" (1929). Still he felt that it was the only solution. Rutherford was cautious. "I have heard that you [Bohr] are on the warpath and wanting to upset the Conservation of Energy both microscopically and macroscopically. I will wait and see before expressing an opinion but I always feel 'there are more things in Heaven and Earth than are dreamt of in our philosophy,'" he wrote in a letter (1929). Paul Dirac disagreed (1929): "I should prefer to keep rigorous conservation of energy at all costs."

Bohr made his speculations public in his Faraday Lecture to the British Chemical Society on May 8, 1930. Noting the problem posed by the continuous β-decay energy spectrum, he remarked, "At the present stage of atomic theory, however, we may say that we have no argument, either empirical or theoretical, for upholding the energy principle in the case of β-decay disintegrations, and are even led to complications and difficulties in trying to do so. Of course, a radical departure from this principle would imply strange consequences if such a process could be reversed." One such consequence was that if an electron was absorbed by a nucleus (inverse β-decay), then nonconservation of energy could provide the energy production mechanism in stars. Bohr went on, "I shall not enter further into such speculations and their possible bearing on the much debated question of the source of stellar energy. I have touched upon them here mainly to emphasize that in atomic theory, notwithstanding all the recent progress, we must still be prepared for new surprises."

Wolfgang Pauli thought otherwise. He wrote to Bohr (1929), "I must say that your paper has given me *little* satisfaction. . . . I do *not* exactly mean that this is unpermissible but it is a risky business. . . . Let the stars radiate in peace." Pauli would soon propose his own startling alternative.

This was Pauli's "desperate way out." In 1930 he suggested that a very light, neutral, spin ½ particle was also emitted in β-decay. This saved the conservation laws, because in a three-body decay the energy of the electron was no longer required to be unique. He made his suggestion in a letter (December 14, 1930, quoted in Pais 1986, 315) to Meitner and Geiger in Tübingen:

Dear radioactive ladies and gentleman,

I have come upon a desperate way out regarding . . . the continuous β-spectrum, in order to save the energy law. To wit, the possibility that there could exist in the nucleus electrically neutral particles, which I shall call neutrons, which have spin ½ and satisfy the exclusion principle and which are further distinct from light-quanta in that they do not move with light velocity. The mass of the neutrons should be of the same order of magnitude as the electron mass and in any case not larger than 0.01 times the proton mass.—The continuous β-spectrum would then become understandable from the assumption that in β-decay a neutron is emitted along with the electron, in such a way that the sum of the energies of the neutron and the electron is constant. . . .

For the time being I dare not publish anything about this idea and address myself confidentially first to you, dear radioactive ones, with the question how it would be with the experimental proof of such a neutron, if it were to have the penetrating power equal to or about ten times larger than a γ-ray.

I admit that my way out may not seem very probable *a priori* since one would probably have seen the neutrons a long time ago if they exist. But only he who dares wins, and the seriousness of the situation concerning the continuous β-spectrum is illuminated by my honored predecessor, Mr. Debye, who recently said to me in Brussels: "Oh, it is best not to think about this at all, as with new taxes." One must therefore discuss seriously every road to salvation. —Thus, dear radioactive ones, examine and judge.—Unfortunately I cannot appear personally in Tübingen since a ball which takes place in Zurich the night of the sixth to the seventh of December makes my presence here indispensable. . . .

Your most humble servant, W. Pauli.

Pauli originally called the particle the neutron (not to be confused with the heavy neutral particle that is a constituent of the atomic nucleus, which is still called the neutron and was discovered by Chadwick in 1932). Enrico Fermi (1934a, 1934b) later christened it the neutrino and almost immediately incorporated the putative new particle into a successful theory of β-decay. Although there were some difficulties for the theory during the 1930s, it was ultimately strongly supported by the experimental evidence. That success provided most physicists with sufficient evidence for the neutrino's existence. As Frederick Reines has put it (1982b), "It must be recognized, however, that independent of the observation of a 'free neutrino' interaction with matter, the [Fermi] theory was so attractive in its explanation of beta decay that belief in the neutrino as a 'real' entity was general." This is another example of the view that to have good reason for holding a theory is, ipso facto, to have good reason for believing in the existence of the entities postulated by that theory.

Finally, in 1956 Reines and Clyde Cowan delivered the first direct evidence of the neutrino's reality (Cowan et al. 1956). The poltergeist, as Reines called it, had finally been observed.

(8)

How Many Neutrinos?

The story of the search for the neutrino begins with Enrico Fermi's theory of β-decay (1934a, 1934b), which incorporated this newly hypothesized particle. The success of this theory quite naturally provided physicists with good reasons to believe in the existence of the neutrino. Although this theory did not enjoy an unbroken success, by the 1950s it was triumphant, well supported by experimental evidence.

One scientist said, "Fermi's theory is remarkable in that it accounts for all the observed properties of β-decay. It correctly predicts the dependence of the radioactive nucleus lifetime on the energy released in the decay. It also predicts the correct shape of the energy spectrum of the emitted electrons. Its success was taken as convincing evidence that a neutrino is indeed created simultaneously with an electron every time a nucleus disintegrates through β-decay" (quoted in Franklin 2000).

The Fermi theory was so successful in the explanation of most of the important features of β-decay that most physicists accepted the neutrino as one of the "particles" of modern physics.

For more details of this episode, see Franklin (2000, chapters 5, 6, and 7).

Nevertheless, some physicists wanted more direct evidence of the neutrino's existence. A typical statement of this view was given by H. Richard Crane in his 1948 review article on the neutrino. "Not everyone would be willing to say that he believes in the existence of the neutrino, but it is safe to say there is hardly one of us who is not served by the neutrino hypothesis as an aid in thinking about the beta-decay process. . . . While the hypothesis has had great usefulness, it should be kept in the back of one's mind that it has not cleared up the basic mystery, and that such will continue to be the case until the neutrino is somehow caught at a distance from the emitting nucleus."

Finding the Poltergeist

In 1950 Enrico Fermi suggested an experiment that he believed would definitively show the existence of the neutrino. "Perhaps the most conclusive proof for the existence of the neutrino, and the most remote of attainment, would be to observe β-decay with recoil of the nucleus and momentum of the electron known so as to give the direction of the neutrino, and then on the path of the neutrino to detect almost simultaneously an inverse β-reaction [$\bar{\nu} + p \rightarrow n + e^+$, antineutrino + proton $\rightarrow$ neutron + positron] whose energy relations agree with the energy of the neutrino emitted in the first reaction" (Fermi 1950, 85).

At least part of Fermi's suggestion would prove quite feasible. Frederick Reines and Clyde Cowan and their collaborators would observe the inverse β-process and provide direct evidence for the existence of the neutrino (Cowan et al. 1956; Reines et al. 1960). However, as Bethe and Bacher (1936) had pointed out soon after Fermi's formulation of his theory, this reaction would be very difficult to observe.

> There is thus considerable evidence for the neutrino hypothesis. Unfortunately, all this evidence is indirect; and more unfortunately, there seems at present to be no way of getting any direct evidence. At least, it seems practically impossible to detect neutrinos in the *free state*, i.e., *after* they have been emitted by the radioactive atom. There is only *one* process which neutrinos can *certainly* cause. That is the inverse β-process, consisting of the capture of a neutrino by a nucleus together

with the emission of an electron (or positron). This process is so extremely rare that a neutrino has to go, in the average, through 10^{16} km of solid matter before it causes such a process [10^{16} km is 1,000 light-years]. The present methods of detection must be improved at least by a factor of 10^{13} [10 million million] in sensitivity before such a process could be detected.

In a prescient comment, Crane (1948) noted a possibility for improving the experiment: "The use of the large neutrino flux from a chain-reacting pile to test for the inverse beta-decay process has been the subject of conversation among physicists since the advent of the pile, and it would be surprising if experiments of this sort were not going forward at the present time in one or more of the government laboratories." The first operating pile, or nuclear reactor, had been constructed in Chicago during World War II by Enrico Fermi and others as part of the effort to build the atomic bomb. A nuclear reactor uses the same nuclear fission process as does the atomic bomb, but under controlled conditions and much more slowly. It produces the same radioactive fission products, which in turn produce antineutrinos. (The antineutrino is the antiparticle of the neutrino.)

Detecting such a rare interaction as inverse β-decay in a reasonable time would require both a large amount of matter as a target and a very large flux of neutrinos. Developments during World War II had made such an experiment feasible. The first experimental suggestion of what became known as Project Poltergeist—so named because of the properties of the elusive neutrino—was to use an atomic bomb, not a nuclear reactor, as the source of antineutrinos. The explosion of an atomic bomb, a nuclear fission weapon, produces many short-lived fission fragments, each of which is radioactive. These fission fragments emit an electron and an antineutrino. The large number of antineutrinos produced could be used to search for inverse β-decay. Reines and Cowan, scientists at the Los Alamos weapons laboratory, where the atomic bomb was first developed and built, hypothesized that they could place a neutrino detector close enough to an atomic bomb explosion so that they would have a good chance of detecting that process.

Certain organic liquids had been found to emit light, or scintillate, when an electron passed through them. This light could then be detected by a photomultiplier tube, a device in which a photosensitive surface emitted electrons when struck by light. These electrons were then multiplied in the tube so that they produced a detectable electronic pulse. The positron emitted by inverse β-decay quickly annihilated with an electron in the liquid, producing two γ-rays. (When matter and anti-matter interact, the particle and its antiparticle annihilate to produce energy.) These γ-rays each had equal energy and were emitted in opposite directions. The γ-rays produced a cascade in which they first ionized an atom, freeing an electron. These electrons in turn produced more γ-rays, which produced more electrons, and so on. This cascade produced detectable light in the liquid scintillator. Reines and Cowan's first plan involved detecting only the γ-rays resulting from the annihilation of the positron produced with an electron. The positron would be evidence that inverse β-decay had occurred and that the neutrino was present. No other known reactions were expected to produce such positrons in any significant numbers.

Detecting the positron was relatively easy, but placing such a detector close enough to a nuclear explosion without its being destroyed was more difficult. Cowan described the planned experiment (1964):

We would dig a shaft near "ground zero" [the nuclear explosion] about 10 feet in diameter and about 150 feet deep. We would put a tank, 10 feet in diameter and about 75 feet long on end at the bottom of the shaft. We would then suspend our detector from the top of the tank, along with its recording apparatus, and back-fill the shaft above the tank.

As the time for the explosion approached, we would start vacuum pumps and evacuate the tank as highly as possible. Then, when the countdown reached "zero," we would break the suspension with a small explosive, allowing the detector to fall freely in the vacuum. For about 2 seconds, the falling detector would be seeing antineutrinos and recording the pulses from them while the earth shock passes harmlessly by, rattling the tank mightily but not disturbing our falling detector. When all was relatively quiet, the detector would reach the bottom of the tank, landing on a thick pile of foam rubber and feathers [figure 8.1].

We would return to the site of the shaft in a few days (when the surface radioactivity had died away sufficiently) and dig down to the tank, recover the detector, and know the truth about neutrinos!

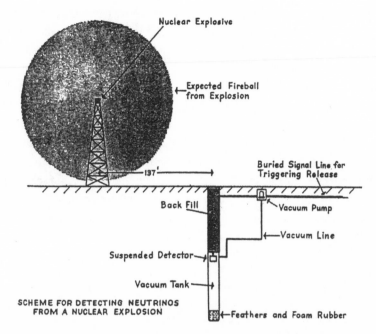

Nuclear Explosive

Expected Fireball
from Explosion

Buried Signal Line for
Triggering Release

137'

Back Fill

Vacuum Pump

Vacuum Line

Suspended Detector

Vacuum Tank

SCHEME FOR DETECTING NEUTRINOS
FROM A NUCLEAR EXPLOSION

Feathers and Foam Rubber

FIGURE 8.1. Reines and Cowan's suggested scheme for detecting neutrinos from a nuclear explosion. The suspended detector is 137 feet from the target. From Cowan (1964).

Cowan further remarked, "As it made little difference precisely where we placed our shaft, we chose to put it 137 feet from the base of the tower for luck." This is an example of physicist humor: 137 is approximately $1/\alpha$, where α is the fine-structure constant, an important quantity in atomic physics. Physicists have no explanation of why α has that particular value. There is a joke that circulates within the physics community. When Wolfgang Pauli, the physicist who first suggested the neutrino and who was notorious for his ego, died, he was welcomed into heaven by God. God asked if there was anything about physics that Pauli wanted to know. Pauli replied that he had never understood why α, the fine-structure constant, was approximately $1/137$. God remarked that he was glad that Pauli had asked and gave Pauli a paper that he (God) had recently written on the subject. Pauli scanned the paper and remarked, "This is wrong."

The rather implausible bomb experiment was encouraged by both Fermi and Norris Bradbury, the director of the Los Alamos laboratory. Bradbury gave Reines and Cowan permission to proceed with planning the experiment. Not too surprisingly, the experiment was never performed. In the fall of 1952, J. M. B. Kellogg, the director of the Physics Division at Los Alamos, suggested that Reines and Cowan review the experimental plan to see if the flux of antineutrinos from a nuclear fission reactor could be used in place of that from an atomic bomb. Although the flux from a reactor was thousands of times lower than that expected from a 50-kton atomic bomb, such an experiment could run for months or a year, in contrast to the 1 or 2 sec expected for a bomb experiment. In addition, Reines and Cowan devised a new method for detecting the antineutrino, which involved the detection of both the positron and the neutron produced in β-decay. They described it as follows:

> The detection scheme is shown schematically in Figure [8.2]. An antineutrino (ν) from the fission products in a powerful production reactor is incident on a water target in which $CdCl_2$ has been dissolved. By reaction (1) [$\nu + p \rightarrow n + e^+$], the incident ν produces a positron (e^+) and a neutron (n). The positron slows down and annihilates with an electron, and the resulting two 0.5 MeV [million electron volts] annihilation gamma rays penetrate the target and are detected in prompt coincidence by the two large scintillation detectors placed on opposite sides of the target. The neutron is moderated by the water and then captured by the cadmium in a time dependent on the cadmium concentration (in our experiment practically all the neutrons are captured within 10 μsec of their production). The multiple cadmium-capture gamma rays are detected in prompt coincidence by the two scintillation detectors, yielding a characteristic delayed-coincidence count with the preceding e^+ gammas. (Reines et al. 1960)

By detecting both the positron and the neutron, the experimenters would demonstrate that inverse β-decay had occurred, which could only have been caused by a neutrino. (No other processes produced those particles.) In principle this is a straightforward measurement. The apparatus is turned on and the number of delayed coincidences is counted. In practice things are very different. In almost every experiment backgrounds are present—events or effects that mimic or mask the desired effect. This is a particularly serious problem in a search

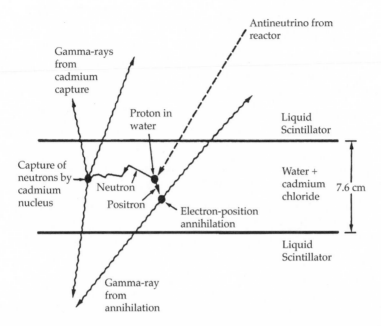

FIGURE 8.2. Schematic diagram of the antineutrino experiment of Reines and Cowan. The γ-rays from both the positron annihilation and the neutron capture by cadmium are detected by the liquid scintillator. From Reines et al. (1960).

for very rare events, such as inverse β-decay. When the apparatus was turned on in their initial experiment, the researchers discovered a large delayed-coincidence background of approximately 5 counts/min. This was not only far larger than the expected signal for inverse β-decay, but also independent of whether the nuclear reactor was on or off. The observed events were not due to antineutrinos, or any other particles, coming from the reactor. They were caused by cosmic rays, particles passing through the apparatus from the atmosphere. Steps were taken to reduce the background.

Reines, Cowan, and their collaborators improved their apparatus and performed a second search for the free antineutrino in an experiment at the Savannah River nuclear reactor (Cowan et al. 1956; Reines et al. 1960). The new experiment had a much larger signal-to-background ratio, and careful arguments were presented to show that the signal was due to inverse β-decay caused by antineutrinos from the reactor.

The experimenters found a clear signal that depended on the reactor power. The signal was far larger with the reactor on than with the reactor off. "For the first part of the series the total net rate (reactor on minus reactor off) was 1.23 ± 0.24 hr^{-1}. For the second part (during which small changes were made) the net rate was 0.93 ± 0.22. From these data we conclude that there was a reactor-associated signal."

On June 14, 1956, after the completion of the experiment, Reines and Cowan sent a telegram to Pauli that said, "We are happy to inform you that we have definitely detected neutrinos from fission fragments by observing inverse beta decay of protons." Reines (1982a) also reported that he had later learned that Pauli and some friends had consumed a case of champagne in celebration of the news. The poltergeist had been found.

How Many Neutrinos?

Following the Reines-Cowan experiment, physicists in the late 1950s faced another problem with the neutrino. This was the question of whether only one kind of neutrino existed. Possibly one neutrino was associated with the electrons in β-decay, an electron neutrino, and another with the muon, a muon neutrino. (The pion decays into a muon and a neutrino; the muon decays into an electron, a neutrino, and an antineutrino.) As Reines later remarked, "Having detected a neutrino associated with nuclear beta decay we puzzled as to whether the neutral particle from (π,μ) decay, was the same as the neutrino from nuclear beta decay" (1982b). Most physicists at the time thought that the idea of two neutrinos introduced an unneeded complexity.

The question of the identity of the muon and electron neutrinos was made more pressing by the failure to observe the decay $\mu \rightarrow e + \gamma$. Ordinary muon decay was thought to be $\mu \rightarrow e + \nu + \bar{\nu}$. If there was only one kind of neutrino, then the neutrino and the antineutrino could annihilate one another before the decay, resulting in the decay $\mu \rightarrow e + \gamma$. The ratio $R = (\mu \rightarrow e + \gamma)/(\mu \rightarrow e + \nu + \bar{\nu})$ could be calculated. At the beginning of 1962, the measured value of R was $< 2 \times 10^{-6}$, in disagreement with the most plausible theoretical estimate of 10^{-4}. In a further study, David Bartlett and collaborators remarked,

"Suggestions have been advanced for a new selection rule or conservation law to explain the absence of $\mu \rightarrow e + \gamma$ decays. . . . Conservation laws have been proposed, involving two sorts of neutrinos, one associated with electrons, the other with muons. Additional support for *such a radical interpretation* would be provided by a still smaller experimental limit on the $\mu \rightarrow e + \gamma$ process" (Bartlett, Devons, and Sachs 1962; emphasis added). The limit they provided was $R < 6 \times 10^{-8}$. In an adjoining paper, Frankel et al. (1962) found a value of $R < 1.9 \times 10^{-7}$. Both of these results were far lower than any existing theoretical estimates. There was clearly a problem. One solution was two neutrinos, an electron neutrino and a muon neutrino. If that was correct, then the neutrino and the antineutrino in muon decay would be different particles—one an electron neutrino and the other a muon neutrino— and they therefore could not annihilate one another, because only particles and their own antiparticles can annihilate one another. The decay $\mu \rightarrow e + \gamma$ would be forbidden.

Discovery of the Muon Neutrino

The question of how many neutrinos existed was intriguing, but its experimental answer would prove difficult. To test the two-neutrino hypothesis, high-energy neutrinos were needed. The energy required was far higher than that of neutrinos from nuclear reactors.

A solution to the technical problem of how to create a beam of high-energy neutrinos with sufficient intensity was proposed independently by Bruno Pontecorvo and by Melvin Schwartz in 1960. Both proposed using high-energy pions produced in the collision of high-energy accelerator protons with a metal target. The pions would then decay into a muon and a neutrino $[\pi^{\pm} \rightarrow \mu^{\pm} + (\nu \,|\, \bar{\nu})]$. The neutrino energy would be a significant fraction of the pion energy and would be emitted along the pion direction. Schwartz remarked,

> That night it came to me. It was incredibly simple. All one had to do was use neutrinos [to study high-energy weak interactions]. The neutrinos would come from pion decay and a large shield could be constructed to remove all background consisting of the strongly and electromagnetically interacting particles and allow only neutrinos through. . . . They

[T. D. Lee and C. N. Yang] also pointed out that this experiment could resolve the long standing puzzle of the missing decay of the muon into electron and gamma. There were clear-cut theoretical predictions, in contradiction to the experiments, that $\mu \rightarrow e + \gamma$ should take place in one in every 10^5 muon decays, unless there is a new quantum number to forbid it. Indeed it became increasingly clear that the only way in which this absence could be explained required that there be two neutrinos, one associated with the electron and the other associated with the muon. In this case, making use of neutrinos from the decay $(\pi \rightarrow \mu + \nu)$ we would only see muons produced, never electrons. Estimates at that point indicated that with 10 tons of detector we might obtain an event per day, if the new Alternate Gradient Synchrotron [a high-energy particle accelerator] at Brookhaven accelerated as much as 10^{11} protons per second. However, the accelerator was still two years from completion and the subject seemed almost academic.

The wait was not in vain. In fact, the kind of beam proposed by Pontecorvo and Schwartz was used by Schwartz, Leon Lederman, Jack Steinberger, and their collaborators in an experiment to test the two-neutrino hypothesis at the Alternate Gradient Synchrotron. (Schwartz, Lederman, and Steinberger would win the Nobel Prize for their work.) Pions and other particles were produced by the collision of high-energy protons with a metal target. The entire flux of particles produced struck 13.5 m of steel in front of a 10-tn spark chamber that served as a detector. (The steel came from the armor plate of a battleship. Although it wasn't beating swords into plowshares, it was cutting armor into shielding, a peaceful use. Other high-energy physics experiments used barrels from naval guns, cutting cannons into collimators.) The shielding removed virtually all of the beam particles except neutrinos. The group obtained a total of 113 pictures; 34 contained single tracks (figure 8.3), which, if interpreted as muons, had momenta greater than 300 MeV/c; 22 were "vertex" events, which had more than one track; and 8 were "showers," which were "in general single tracks, too irregular in structure to be typical of μ mesons, and more typical of electron or photon showers" (Danby et al. 1962).

The experimenters argued that the observed events were not produced by the possible backgrounds of cosmic rays or neutrons. The cosmic ray background was estimated by operating the experimental

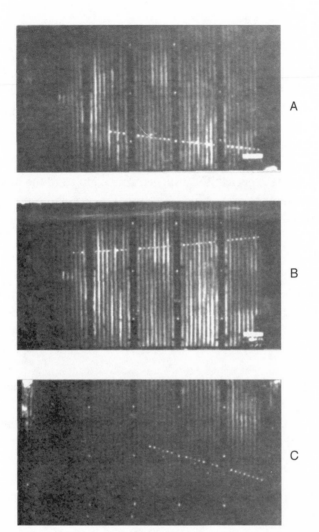

FIGURE 8.3. Single muon events, with long straight tracks. From Danby et al (1962).

apparatus with the accelerator off. The researchers found a total of 5 ± 1 events that could be attributed to cosmic rays.

The experimenters also argued that the single particles observed were muons and that they resulted from neutrino interactions. They noted that the single tracks traversed a total of 820 cm of aluminum without producing a single "clear" nuclear interaction. The interaction length for 400-MeV pions, the alternative explanation of the observed tracks, was less than 100 cm. "We should, therefore, have observed of the order of 8 'clear' interactions [if the tracks were pions]; instead we observed none," they wrote. The tracks were muons, not pions.

It was clear that neutrinos were producing muons. The question remained whether they were also producing electrons. "Are there two kinds of neutrinos? The earlier discussion leads us to ask if the reactions [that produce muons and electrons] occur with the same rate. This would be expected if ν_μ, the neutrino coupled to the muon and produced in pion decay, is the same as ν_e, the neutrino coupled to the electron and produced in nuclear beta decay." The collaborators noted that the tracks for their muon events [figure 8.3] were quite different from the showers produced in their spark chambers by 400-MeV electrons [figure 8.4]: "We have observed 34 single muon events of which 5 are considered to be cosmic-ray background. If $\nu_\mu = \nu_e$, there should be of the order of 29 electron showers with a mean energy greater than 400 MeV/c. Instead, the only candidates which we have for such events are six 'showers' of qualitatively different appearance from those of Fig. [8.4]."

The experimenters concluded, "The most plausible explanation for the absence of the electron showers . . . is that $\nu_\mu \neq \nu_e$; that there are at least two types of neutrinos. This also resolves the problem raised by the forbiddenness of the $\mu^+ \rightarrow e^+ + \gamma$ decay." The discovery of the muon neutrino led physicists to formulate two separate conservation laws, one for electron family members and one for muon family members. Previously theory required only that the number of leptons, or light particles, be conserved. Now the decay of the muon was characterized as $\mu^- \rightarrow e^- + \bar{\nu}_e + \nu_\mu$. This conserved both family numbers. If ν_e is not the antiparticle of ν_μ, then there can be no decay $\mu \rightarrow e + \gamma$. Now there were two.

Discovery of the τ-Lepton and Its Neutrino

The muon and its neutrino had already added complexity to the system of elementary particles. The muon seemed, in all respects, to be merely a heavy electron. As I. I. Rabi asked, "Who ordered that?" The

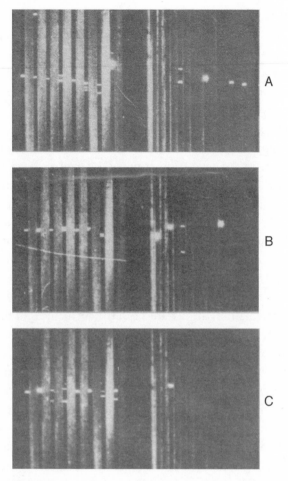

FIGURE 8.4. The tracks of 400-MeV electrons. These produce showers and are quite different from the muon tracks. From Danby et al (1962).

situation became even more complex when Martin Perl, who would win the Nobel Prize for this discovery, and his collaborators found evidence for a third lepton, the τ (Perl et al. 1975).

In an experiment at SPEAR, the electron-positron collider at the Stanford Linear Accelerator Center (SLAC), a SLAC-Berkeley collaboration had found evidence for events with electrons and muons of opposite charge. After demonstrating that the observed events could not be caused by other sources, the group concluded "that the signature e-μ events cannot be explained either by the production and decay of any presently known particles or as coming from any well-understood interactions which can conventionally lead to an e and a μ in the final state. *A possible explanation is the production and decay of a pair of new particles, each having a mass in the range 1.6 to 2.0 GeV/c²"* (Perl et al. 1975). Further work by these collaborators refined their result (Perl et al. 1976, 1977).

The later experiments showed that the e-μ events were produced by the pair production of a new lepton, the τ, which decayed into an electron or muon and two neutrinos. One of the neutrinos was associated with the electron or the muon by the conservation of electron family number or muon family number, and the other was a neutrino associated with the τ, v_τ. The τ-lepton had been discovered, and it was inferred that there was an associated neutrino. The collaborators also argued persuasively that the decay of the charged τ-lepton contained its own associated neutrino, v_τ. They noted that their calculation of various measured properties of the τ required such a neutrino. The success of the calculations argued for both the correctness of their model and the existence of the τ-neutrino.

A group at Fermilab recently reported more direct, albeit only suggestive, evidence for the τ-neutrino (unpublished data). Protons from the Tevatron, a high-energy accelerator at Fermilab, struck a tungsten target. In those high-energy collisions, many particles were produced, including τ-leptons, which subsequently decayed, producing τ-neutrinos. Shielding eliminated all particles except neutrinos, which interact very weakly with matter. The neutrinos emerging from the shielding struck an advanced emulsion, similar to photographic film. Among the particles detected in the emulsion were four events with a kinked track characteristic of τ-lepton decay. The experimenters argued that such τ-leptons could only be produced by τ-neutrinos.

More evidence has since been found on the existence of the τ-neutrino. In addition to the data provided by the SLAC-Berkeley collaboration, there is extremely persuasive evidence that there are, in fact, three—and only three—neutrinos.

Measurements on the Z^0 boson, an elementary particle, provide substantiation. The Z^0 has a lifetime (it decays), a mass, and an intrinsic width. (This means that even if the mass of the particle is measured with a perfect apparatus, the resulting mass will have a range of values.)

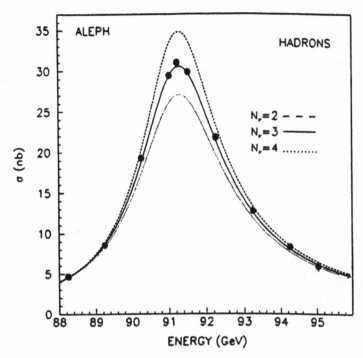

FIGURE 8.5. The measurement of the width of the Z^0 particle. The predictions for the number of neutrinos, N = 2, 3, and 4, are shown. From Decamp et al. (1990).

The Heisenberg uncertainty principle posits that if a particle has a finite lifetime, then it also must have intrinsic width. The shorter the lifetime of the particle, the larger the width. The lifetime, and thus the width, depend on the number of ways in which a particle can decay. The more decay modes there are, the shorter the lifetime is. Physicists have accurately measured the width of the Z^0. This can then be compared to predictions of that width for varying numbers of neutrinos. (The Z^0 can decay into pairs of different neutrinos, ν_e-$\bar{\nu}_e$, ν_μ-$\bar{\nu}_\mu$, ν_τ-$\bar{\nu}_\tau$, etc. The more kinds neutrinos there are, the shorter the lifetime and the larger the width.) The production of Z^0s as a function of energy is shown in figure 8.5. N_ν = 3 is clearly favored, thus providing evidence for the τ-neutrino, in addition to the electron and muon neutrinos.

Now there are three.

(9)

The Appearance and Disappearance
of the 17-keV Neutrino

One of the properties of the neutrino, the third particle discovered to be emitted in β-decay, was believed to be that it had zero mass (the best current experiments give the neutrino a very small mass). For a time, physicists believed a heavy neutrino might exist—the 17-keV (thousand electron volts) neutrino—which would be emitted in place of the ordinary neutrino in a fraction of all β-decays. Experimental results both for and against the existence of the 17-keV neutrino were reported by various groups. Both the original proposal, and all subsequent positive claims, were obtained in experiments using one type of apparatus, which incorporated a solid-state detector, whereas the initial negative evidence resulted from experiments using another type of detector, a magnetic spectrometer. These were both seemingly reliable types of experimental apparatus. Solid-state detectors had been widely employed since the early 1960s, and their use was well understood. Magnetic spectrometers had been used in nuclear β-decay experiments

For more details of this episode, see Franklin (1995a).

since the 1930s, and both the problems and advantages of this technique had been well studied. It was possible that the discordant results were due to some crucial difference between the types of apparatus or to different sources of background that might mimic or mask the signal. Questions were also raised about the proper theoretical model with which to compare the experimental data and the appropriate energy range for that comparison. Eventually these problems were solved, and the conclusion was reached that the 17-keV neutrino did not exist.

The Appearance

The 17-keV neutrino was "discovered" by John Simpson in 1985. He had searched for a heavy neutrino by looking for a kink in the decay energy spectrum, or in the Kurie plot, at an energy equal to the maximum allowed decay energy minus the mass of the heavy neutrino, in energy units. (The Kurie plot is a particular mathematical function of β-decay quantities plotted as a function of the decay electron energy. It has the nice visual property that the Kurie plot for the correct theory is a straight line.) In figure 9.1, the dotted lines give the Kurie plots for both a heavy neutrino (lower line) and the ordinary light neutrino (upper line). The heavy-neutrino line has a smaller energy intercept because the mass of the heavy neutrino reduces the maximum allowed energy for the decay electron. (In this example the mass of the heavy neutrino is two energy units.) If, as Simpson suspected, both types of neutrino were emitted in β-decay, then the Kurie plot for the decay would be the sum of the Kurie plots for each of the neutrinos (solid line). The figure shows a distinct kink, a change in the slope of the line, at an energy equal to the maximum allowed energy minus the mass of the neutrino. Simpson, and others, expected that the heavy-neutrino emission would be only a small fraction of that of the ordinary neutrino. This would make the kink even more difficult to observe than in the case illustrated in the figure.

Simpson actually examined the fractional change in the Kurie plot, $\Delta K/K$, which is more sensitive to the presence of another neutrino. His initial experimental result for the decay energy spectrum of tritium is

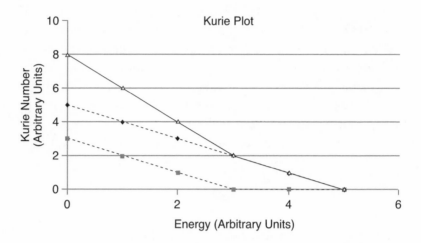

FIGURE 9.1. The dotted lines give the Kurie plots for both a heavy neutrino (lower) and the ordinary light neutrino (upper). The heavy-neutrino line has a smaller energy intercept because the mass of the heavy neutrino reduces the maximum allowed energy for the decay electron. (In this example the mass of the heavy neutrino is two energy units.) The solid line is the sum of the two plots. A kink, a change in slope, is clearly visible at the maximum energy minus the mass of the heavy neutrino.

shown in figure 9.2. A kink, a marked change in slope of the $\Delta K/K$ graph, is clearly seen at an electron energy, T_β, of 1.5 keV, corresponding to a 17-keV neutrino. (The maximum decay energy for tritium is 18.6 keV. If there were no effect from the presence of a heavy neutrino, this graph would be a horizontal straight line.) "In summary, the β spectrum of tritium recorded in the present experiment is consistent with the emission of a heavy neutrino of mass about 17.1 keV and a mixing probability [the fraction of heavy neutrinos] of about 3%" (Simpson 1985). Later work, including some by Simpson, reduced the size of the positive effect to approximately 1%.

Within a year there were five attempted replications of Simpson's experiment (Altzitzoglou et al. 1985; Apalikov et al. 1985; Datar et al. 1985; Markey and Boehm 1985; Ohi et al. 1985). Each was negative. The experiments set limits of less than 1% for a 17-keV neutrino branch of the decay, in contrast to Simpson's value of 3%. A typical result, that

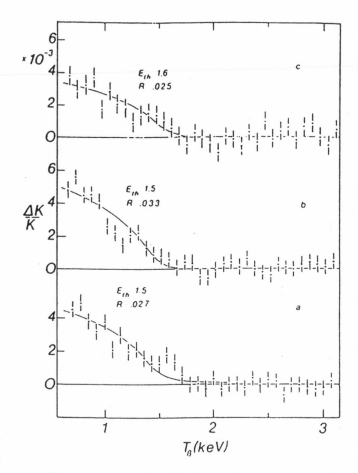

FIGURE 9.2. Simpson's results for $\Delta K/K$ (the fractional change in the Kurie plot) as a function of the kinetic energy of the β-particles. E_{th} is the threshold energy, the difference between the end-point energy and the mass of the heavy neutrino. A kink is clearly seen at $E_{th} = 1.5$ keV, or at a mass of 17.1 keV. From Simpson (1985).

of T. Ohi and others, is shown in figure 9.3. No kink of any kind is apparent.

Each of the subsequent experiments had examined the β-decay spectrum of ^{35}S rather than that of tritium ($^3H_1$). Three of the experiments used magnetic spectrometers. The other two used solid-state detectors, the same type used by Simpson. In the latter two cases, how-

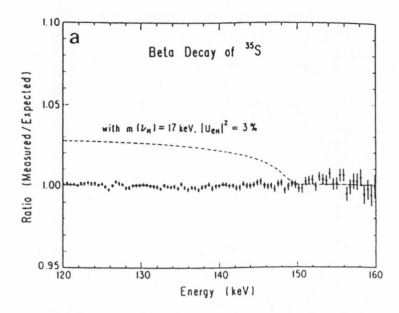

FIGURE 9.3. The ratio of the measured ^{35}S electron energy spectrum to the theoretical spectrum. A 3% mixing of a 17-keV neutrino should distort the spectrum as indicated by the dashed curve. From Ohi et al. (1985).

ever, the source was not implanted in the detector, as Simpson's had been, but was separated from it. Such an arrangement would change the atomic physics corrections to the spectrum. (To compare the measured spectrum to the predicted spectrum, the experimenters needed to make atomic physics corrections because of the interaction of the decay electrons with the atomic electrons.) The ^{35}S β-decay sources used in the experiments had a higher end-point energy than did the tritium used by Simpson, which made the atomic physics corrections to the β-decay spectrum less important.

Simpson's first report of the 17-keV neutrino was unexpected. It was not predicted, or even suggested, by any existing theory. Faced with such an unexpected result the physics community took a reasonable approach. Some theoretical physicists tried to explain the result within the context of accepted theory. They argued that a plausible alternative explanation of the result had not been considered. This involved whether the correct theory had been used in analyzing the data and comparing the experimental result with the theory of the phenomenon. This is an important point. An experimental result is not usually immediately established by examining the raw data; considerable analysis is required. In this case the analysis included atomic physics corrections, needed for the comparison between the theoretical spectrum and the experimental data. Everyone involved agreed that such corrections were needed; the disagreement involved what the proper corrections were. Several calculations by others indicated, at least qualitatively, that Simpson's result could be accommodated within accepted theory and that there was no need for a new particle. "A detailed account of the decay energy and Coulomb-screening effects raises the theoretical curve in precisely this energy range so that little, if any, of the excess remains" (Lindhard and Hansen 1986).

The combination of negative experimental searches combined with plausible theoretical explanations of Simpson's result had a chilling effect on the field. Almost all experimental work on the subject ceased. Simpson, however, continued his work. He presented further evidence in support of the 17-keV neutrino, obtained with a somewhat modified experimental apparatus (1986a, 1986b). He also took the criticism of his work seriously and presented an analysis of his new data that incorporated the atomic physics corrections suggested by his critics. Although this reduced the size of his effect by approximately 20%, the effect was still clearly present. Simpson had shown that his result was reasonably robust under variations in the atomic physics corrections to the decay spectrum. He also questioned whether or not the analysis procedures used in the five negative searches were adequate to set the upper limits they had reported. He argued that the wide energy range used by his critics to fit the β-decay spectrum tended to minimize any possible effect of a heavy neutrino, which would appear primarily in a narrow energy band near the maximum energy allowed for the heavy neutrino. Simpson further questioned whether or not the "shape-correction" factor needed to fit the spectra in magnetic spectrometer experiments could mask a kink caused by a heavy neutrino. It is generally quite difficult to calculate the efficiency for magnetic spectrometer apparatuses. To get a good fit to the observed spectrum a shape-

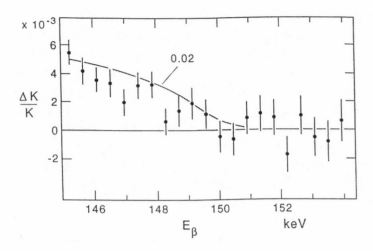

FIGURE 9.4. ΔK/K for the ³⁵S spectra of Ohi and others as recalculated by Simpson. A kink is clearly visible. From Simpson (1986b).

correction factor (some might say a fudge factor) is used. These questions would have to be answered.

Simpson also presented a reanalysis of Ohi's data for which he used his own preferred analysis procedure (1986b). He found a positive effect (figure 9.4). How the same data could both support and refute the existence of the 17-keV neutrino was a problem that needed resolution.

Further negative evidence was provided by other experiments. D. Hetherington and others, for example, urged caution concerning Simpson's method of data analysis. They pointed out that "concentrating on too narrow a region [of energy] can lead to misinterpretation of a local statistical anomaly as a more general trend" (Hetherington et al. 1987). The issue of the appropriate energy range for the analysis of the data had not yet been decided. At the end of 1988, the situation was much as it was at the end of 1985. Simpson had presented positive results on a 17-keV neutrino. There were nine negative experimental reports as well as plausible theoretical explanations of his result and no strong evidence for the existence of the 17-keV neutrino.

In 1989 Simpson and Andrew Hime presented two additional positive results, with both tritium and ³⁵S, the spectrum used in the original negative searches. In these reports the value of the mixing fraction of 17-keV neutrinos had been reduced to approximately 1% (Hime and Simpson 1989; Simpson and Hime 1989).

Dramatic changes came in 1991. New positive results were reported by groups at Oxford and at Berkeley (Hime and Jelley 1991; Sur et al. 1991). These results were quite persuasive. Hime and Nicholas Jelley (Oxford) had improved the experimental apparatus to eliminate a problem found in earlier experiments. The Berkeley group took precautions to ensure that the decay electrons deposited their full energy in their detector. If the full energy wasn't deposited, this would distort the decay energy spectrum and simulate the kink that signaled the presence of a heavy neutrino. The Berkeley group claimed that their result "supports the claim by Simpson that there is a 17-keV neutrino emitted with ~1% probability in β decay." These results generated considerable new experimental and theoretical work. Sheldon Glashow, a Nobel Prize–winning theorist, remarked (1991), "Simpson's extraordinary finding proves that Nature's bag of tricks is not empty, and demonstrates the virtue of consulting her, not her prophets." Glashow was advocating the primacy of experimental evidence over theoretical speculation and calculation. Unfortunately, the positive experimental evidence was incorrect.

The Disappearance

The summer of 1991 marked the high point in the life of the 17-keV neutrino. From this time forward only negative results would be reported, and errors would be found in the most persuasive positive results. Leo Piilonen and Alexander Abashian (1992) suggested that Hime and Jelley had overlooked a background effect that might have simulated the effect of a 17-keV neutrino in their experiment. The appearance of several negative results encouraged Hime to consider the Piilonen-Abashian suggestion seriously and to reanalyze his own result. He used an experimentally checked calculation and discovered that he could explain his own Oxford result (with Jelley) without the need for a 17-keV neutrino. "It will be shown that scattering effects are sufficient to describe the Oxford β-decay measurements and that the model can be verified using existing calibration data" (1993). He also suggested that similar effects might explain his earlier positive results obtained in collaboration with Simpson.

Hime briefly reviewed the evidential situation, noting that the major evidence against the existence of the 17-keV neutrino came from magnetic spectrometer experiments in which questions had been raised concerning the shape corrections. He commented that Giovanni Bonvicini (1993) had shown that experimental difficulties combined with limited statistics could mask the presence of a heavy neutrino signature and still be described by a smooth shape correction. Bonvicini's work was very important. By showing that a smooth shape-correction factor might either mask or enhance a kink caused by a 17-keV neutrino, he cast considerable doubt on the early negative results obtained with magnetic spectrometers. This work was influential in persuading scientists to perform the later, more stringent, experimental tests. He remarked, however, that "a measurement of the ^{63}Ni spectrum [by the Tokyo group] has circumvented this difficulty. The sufficiently narrow energy interval studied, and the very high statistics accumulated in the region of interest, makes it very unlikely that a 17-keV threshold has been missed in this experiment." He also cited a new result from a group at Argonne National Laboratory (Mortara et al. 1993) that provided "convincing evidence against a 17-keV neutrino." These negative results provided the impetus for Hime's reexamination of his result.

Some of the evidence that Hime cited against the 17-keV neutrino was provided by the Argonne group. This experiment used a solid-state detector (the same type used originally by Simpson) and an external ^{35}S source. The final result for the mixing probability of the 17-keV neutrino, shown in figure 9.5, was -0.0004 ± 0.0008 (statistical uncertainty) ± 0.0008 (systematic uncertainty). This result, less than 0.2%, was far lower than the 1% reported by Simpson and others. No evidence had been found for a 17-keV neutrino.

The researchers also demonstrated the sensitivity of their apparatus to a possible 17-keV neutrino. They mixed a small component of ^{14}C in their ^{35}S source and detected the resulting kink in their composite spectrum. (The two spectra had different end-point energies. Combin-

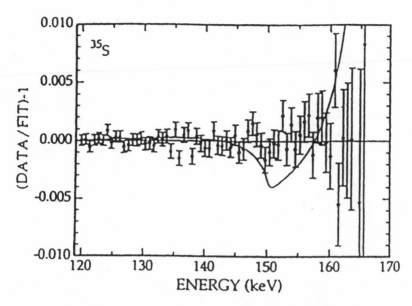

FIGURE 9.5. Residuals (the ratio of the measured result to the expected result minus one) from a fit to the ^{35}S data assuming no massive neutrino. The solid curve represents the residuals expected for decay with a 17-keV neutrino with a mixing probability of 0.85%. No evidence of a 17-keV neutrino is visible. From Mortara et al. (1993).

ing them would result in a composite spectrum with a kink. See figure 9.6) They had shown that their experiment would have detected the 17-keV neutrino had it been present, thus answering a criticism of some of the earlier negative experiments. "*This exercise demonstrates that our method is sensitive to a distortion at the level of the positive experiments. Indeed, the smoother distortion with the composite source is more difficult to detect than the discontinuity expected from the massive neutrino.* In conclusion, we have performed a solid-state counter search for a 17 keV neutrino with an apparatus with demonstrated sensitivity. We find no evidence for a heavy neutrino, in serious conflict with some previous experiments" (Mortara et al. 1993; emphasis added).

At this time the Berkeley collaborators began to question their own positive result. Further experimental runs had shown that their apparatus seemed to generate a spurious 17-keV neutrino signal. They

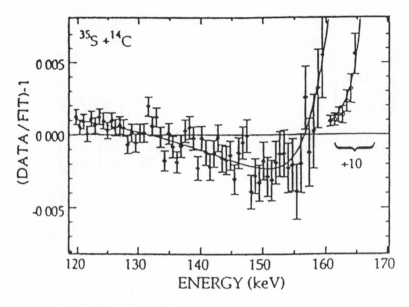

FIGURE 9.6. Residuals from fitting the β-spectrum of a mixed source of ¹⁴C and ³⁵S with a pure ³⁵S shape. The solid curve indicates residuals expected, including the known ¹⁴C contamination. It is a very good fit to the data. From Mortara et al. (1993).

searched for the cause of the artifact. In 1993 they found that, despite their precautions, not all of the electrons were depositing their full energy in their detector, simulating the presence of a 17-keV neutrino.

The newer negative results were persuasive not only because of their improved statistical accuracy, but also because they were able to demonstrate that their experimental apparatuses could detect a kink in the spectrum if one were present. This was a direct experimental check that there were no effects present that would mask the presence of a heavy neutrino. The Tokyo experimenters had also shown that the shape-correction factors used in their experiment did not mask any possible 17-keV neutrino (Kawakami et al. 1992; Ohshima 1993; Ohshima et al. 1993). This combination of almost overwhelming, persuasive evidence against the existence of a 17-keV neutrino, combined with the demonstrated and admitted problems with the positive results, decided the issue. There was no 17-keV neutrino.

This decision was based on experimental evidence, discussion, and criticism—in other words, epistemological criteria. It had been shown that the two most persuasive positive results had overlooked effects that mimicked the presence of a 17-keV neutrino. In addition, the new negative results answered the previous criticisms concerning the shape-correction factor and demonstrated that they could detect a kink in the spectrum if one were present.

Simpson had reanalyzed Ohi's data with strikingly different results and had raised the issue of selectivity. One aspect of this issue was the choice of the energy range used to fit the decay energy spectrum, so that the experimental and theoretical spectra could be compared. Simpson had argued that because 45% of the effect expected occurred within 2 keV of the neutrino threshold, a narrow energy range around that threshold should be used. "In trying to fit a very large portion of the β spectrum, the danger that slowly-varying distortions of a few percent could bury a threshold effect seems to have been disregarded. One cannot emphasize too strongly how delicate is the analysis when searching for a small branch of a heavy neutrino, and how sensitive the result may be to apparently innocuous assumptions" (1986a). Hetherington and others suggested caution.

It has been argued [by Simpson] that in order to avoid systematic errors, only a narrow portion of the beta spectrum should be employed in looking for the threshold effect produced by heavy neutrino mixing. If one accepts this argument, our data in the narrow scan region set an upper limit of 0.44% [much lower than the 3% effect originally found by Simpson]. However, we feel that concentrating on a narrow region and excluding the rest of the data is not warranted provided adequate care is taken to account for systematic errors. The rest of the spectrum plays an essential role in pinning down other parameters such as the endpoint. Furthermore, concentrating on too narrow a region can lead to misinterpretation of a local statistical anomaly as a more general trend which, if extrapolated outside the region, would diverge rapidly from the actual data. (Hetherington et al. 1987)

Hime, one of Simpson's collaborators, agreed (1992). "The difficulty remains, however, that an analysis using such a narrow region could mistake statistical fluctuations as a physical effect. The claim of positive effects in these cases [by Simpson] should be taken lightly without a more rigorous treatment of the data."

This issue was dramatically demonstrated by Simpson's reanalysis of the data of Ohi et al. Figure 9.3 illustrates that Ohi's result showed no evidence for a 17-keV neutrino. Simpson's reanalysis of that same data shows clear positive evidence (figure 9.4). The same data provided both positive and negative evidence for the same effect because the analysis procedures were quite different. Ohi and collaborators had used a wide energy range for their analysis. As Douglas Morrison (1992) showed, the positive effect found by Simpson was due to his use of a narrow energy range for his reanalysis of Ohi's data. (This is a rather technical discussion, but it demonstrates the importance of analysis procedures in producing an experimental result.)

> The question then is, How could the apparently negative evidence of Figure [9.3] become the positive evidence of Figure [9.4]? The explanation is given in Figure [9.7], where a part of the spectrum near 150 keV is enlarged. Dr. Simpson only considered the region 150 keV ± 4 keV (or more exactly +4.1 and −4.9 keV). The procedure was to fit a straight line, shown solid, through the points in the 4 keV interval above 150 keV, and then to make this the base-line by rotating it down through about 20° to make it horizontal. This had the effect of making the points in the interval 4 keV below 150 keV appear above the extrapolated dotted line. This, however, creates some problems, as it appears that a small statistical fluctuation between 151 and 154 keV is being used: the neighboring points between 154 and 167, and below 145 keV, are being neglected although they are many standard deviations away from the fitted line. [Simpson's straight-line fit to the data just above 150 keV and its extrapolation is the line going from lower left to upper right. Comparing the data points to this line generates the positive effect seen in figure 9.7. The dotted curve above the data is the effect expected for a heavy neutrino.] Furthermore, it is important, when analyzing any data, to make sure that the fitted curve passes through the end-point of about 167 keV, which it clearly does not.

The caution urged by both Hetherington and collaborators and by Hime was justified.

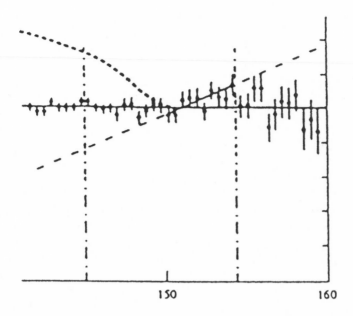

FIGURE 9.7. Morrison's reanalysis of Simpson's reanalysis of Ohi's result. From Morrison (1992).

Several later experiments used both a narrow and a wide energy range in the analysis of their data. Ultimately, the decision that the 17-keV neutrino did not exist was based on finding errors in the two most persuasive positive results and on the overwhelming negative evidence provided by experiments that explicitly avoided the data analysis issues posed by narrow versus wide energy range.

There was no 17-keV neutrino. The kink was dead.

The Missing Solar Neutrinos

I am aware that many critics consider the conditions in stars not
sufficiently extreme to bring about the transmutation—that stars
are not hot enough. . . . We tell them to go and find a *hotter place.*

Arthur Eddington (1927)

The detection of neutrinos by Frederick Reines and Clyde Cowan
opened up a new area of research, in which the neutrino could be used
as a tool to investigate other aspects of the world. One of the most important applications of the new neutrino technology was the investigation of the interior of the sun. The light from the sun cannot be used
for such an investigation because the density of solar matter, combined
with the strength of the electromagnetic interaction, results in an very
large number of absorptions and reemissions of photons as they proceed from the solar interior to its surface. On average, it takes approximately 10 million years for the light emitted by atoms in the center of
the sun to reach the surface. In addition, the photon at the surface is
usually quite different from the one originally emitted.

We can, however, learn about the interior of the sun by examining
the neutrinos emitted in the nuclear reactions that produce the sun's

For more details, see Franklin (2000, chap. 8 and 9). For a very different account, see
Collins and Pinch (1993).

energy. The major source of that energy is the burning of hydrogen (or protons) to produce helium ($4p \rightarrow {}^4He_2 + 2e^+ + 2\nu_e$). This process may proceed through many different nuclear reactions, but the net result is the production of one helium nucleus, two positrons, two electron neutrinos, and 25 MeV (million electron volts) of energy. The number of neutrinos emitted is a sensitive function of the structure of the sun. The fact that neutrinos interact only very weakly with matter makes them an ideal probe of the solar interior. They are essentially unchanged as they proceed from the interior of the sun to a detector on earth.

Davis's Homestake Mine Experiment

The search for solar neutrinos is an illustration of the fruitful interaction of theory and experiment. As John Bahcall and Raymond Davis, the leading physicists involved in the search, remarked, "Theory and observation depend on each other for their significance in solar neutrino research. Without a well-defined predicted counting rate the observed number of captures per day loses most of its meaning. Similarly, the theoretical work derives its motivation from the possibility of experimental tests" (Bahcall and Davis 1989, 488).

The investigation began with the 1946 suggestion by Bruno Pontecorvo that inverse β-decay processes such as neutrino absorption ($\nu + Z \rightarrow e^- + (Z + 1)$ or $\nu + Z \rightarrow e^+ + (Z - 1)$), where Z is a nucleus with charge Z, might be used to provide evidence for the existence of the neutrino. At the time, Pontecorvo, in contrast to most physicists, did not believe that the available evidence provided sufficient support for either the existence of the neutrino or the correctness of Fermi's theory of β-decay. He noted that although the detection of electrons or positrons emitted in these processes would be extremely difficult, the nucleus produced "may be (and generally will be) radioactive. . . . Consequently the radioactivity of the produced nucleus may be looked for as proof of the inverse β-process." He suggested that ${}^{37}Cl$, ${}^{79}Br$, and ${}^{81}Br$ would be suitable targets. "The experiment with chlorine, for example, would consist in irradiating with neutrinos a large volume of chlorine or carbon tetrachloride for a time of order of one month, and extracting the radioactive ^{37}Ar from such a volume by boiling. The radioactive argon would be introduced inside a small counter; the counter efficiency is close to 100 percent." Pontecorvo noted that, based on his estimates, the predicted flux of solar neutrinos was too low for detection and that the energy of the expected solar neutrinos was quite low.

During the 1950s, Davis, who worked at Brookhaven National Laboratory, began building such a detector. (Davis received the 2002 Nobel Prize for his work on solar neutrinos.) He used 3,800 L of carbon tetrachloride (CCl_4), which was buried 19 ft underground to reduce background caused by cosmic rays. He found an upper limit for the flux of solar neutrinos of 40,000 SNU (solar neutrino unit; 1 SNU = 10^{-36} captures per target atom per second). At the time, the theoretically predicted flux was much lower. When Davis submitted his result for publication, a reviewer remarked, rather unkindly, "Any experiment such as this, which does not have the requisite sensitivity, really has no bearing on the existence of neutrinos. To illustrate my point, one would not write a scientific paper describing an experiment in which an experimenter stood on a mountain and reached for the moon, and concluded that the moon was more than eight feet from the top of the mountain." A larger and more sensitive detector was clearly needed.

There was, however, a serious problem. Most of the energy produced in the sun was thought to come from proton-proton interactions, which produced neutrinos with an energy of 0.4 MeV. An energy of 0.86 MeV is required to initiate the ^{37}Cl $\rightarrow$ ^{37}Ar reaction (v_e + ^{37}Cl $\rightarrow$ ^{37}Ar + e$^-$). Solar neutrinos with sufficient energy were those from the carbon-nitrogen-oxygen cycle, which was expected to provide only a small fraction of the sun's energy production, and from the decay of ^{8}B or ^{7}Be produced in the sun. Things did not look promising.

A dramatic improvement came in 1958 when researchers found that production processes in the sun that would produce high-energy neutrinos, high enough to initiate the chlorine reaction, were far more prevalent than theoretically expected. This included production of ^{8}B and ^{7}Be. A further calculation of the expected capture rate of the ^{8}B neutrinos indicated a capture rate of 7.7 per 1,000 gal C_2Cl_4 (perchlorethylene) per day, or 3,900 SNU. This estimate was wildly optimistic. Most of the other measurements and calculations at the time were far more pessimistic.

Despite the pessimistic outlook, work continued. Bahcall and collaborators published the first detailed calculation of solar neutrino fluxes expected (1963). They calculated fluxes that corresponded to a capture rate of 0.01 captures per day in a 1,000-gal detector. "This calculation did not provide any encouragement to build a larger experiment, because even 100,000 gal would only capture one neutrino a day according to this estimate." This calculation included only ^{37}Cl decays to the ground state of ^{37}Ar. The outlook improved considerably when Bahcall showed that the expected capture rate for ^8B neutrinos to excited states of ^{37}Ar was, in fact, 20 times larger than previously expected. This increased the expected capture rate to 4 to 11 events per day in a 100,000-gal detector.

In 1964 Bahcall presented a theoretical analysis of the proposed solar neutrino experiment, and Davis presented a discussion of experimental results already obtained and plans for the future. Bahcall described the motivation for the experiment: "No *direct* evidence for the existence of nuclear reactions in the interior of stars has yet been obtained because the mean free path for photons emitted in the center of a star is typically less than 10^{-10} of the radius of the star. Only neutrinos, with their extremely small interaction cross sections, can enable us to *see into the interior of a star* and thus verify directly the hypothesis of nuclear energy generation in stars."

Using the best available information on solar structure and nuclear reaction cross sections, Bahcall estimated that the number of absorptions per terrestrial ^{37}Cl atom per second would be 40 ± 20 SNU.

In an adjoining paper, Davis (1964) described the experimental situation. He reported the results of an experiment that used the $^{37}Cl \rightarrow {}^{37}Ar$ reaction as a detector. The experiment contained two 500-gal tanks of C_2Cl_4, located 2,300 ft below the surface of the earth in a limestone mine. He calibrated the apparatus by injecting a known amount of argon into the detector and measuring how much was detected when the tank was swept with helium gas. He measured the efficiency to be greater than 95%. He also found an observed counting rate of three events in 18 days, which he attributed to cosmic ray background. He used Bahcall's calculation to determine that the observed counting rate in 100,000 gal C_2Cl_4 would be 4 to 11 per day, "which is an order of magnitude larger than the counter background."

Work on Davis's large neutrino detector began in early 1965. To reduce the background caused by cosmic rays, the apparatus was to be located in a room 4,850 ft underground in the Homestake gold mine in Lead, South Dakota. A cavern 30 × 60 × 32 ft, large enough to contain the 100,000-gal tank to hold the C_2Cl_4, was excavated. During this period there was little theoretical work done on the problem of solar neutrinos. The last theoretical calculations of the expected flux done, by Bahcall and collaborators, before the first experimental results were reported gave results between 8 and 29 SNU and between 7 and 49 SNU.

Davis published the first results from the Homestake detector in 1968. The experimenters described their apparatus (Davis, Harmer, and Hoffman 1968): "A detection system that contains 390,000 liters (520 tons chlorine) of liquid tetrachloroethylene, C_2Cl_4, in a horizontal cylindrical tank was built along the lines proposed earlier. The system is located 4850 ft underground in the Homestake gold mine at Lead, South Dakota. It is essential to place the detector underground to reduce the production of ^{37}Ar from [background reactions]. . . . The total ^{37}Ar production from all background processes is less than 0.2 ^{37}Ar atom per day, which is well below the rate expected from solar neutrinos. [The expected rate was between 4 and 11 events per day.]"

Because the expected amount of ^{37}Ar produced was expected to be small, an efficient method for recovering the argon gas from the C_2Cl_4 was needed. The experimenters circulated helium gas through the liquid in the tank to sweep out the ^{37}Ar atoms produced by neutrinos. They checked the efficiency of their recovery method in two different ways. Both methods gave a recovery efficiency of greater than 95%.

The data from this initial experiment were obtained in two separate experimental runs. The first was an exposure of 48 days with a counting period of 39 days. (The half-life of ^{37}Ar is 35 days, so that more than one-half of the decays would be counted.) "It was counted for 39 days and the total number of counts observed in the ^{37}Ar peak position in the pulse height spectrum was 22 counts. This rate is to be compared with a background rate of 31 ± 10 counts for this period," Davis and colleagues wrote. The background was measured in an identical proportional counter filled with nonradioactive ^{36}Ar. No signal above background had been observed. The second exposure was for 110 days. "It may be seen that from the pulse-height for the first 35 days that 11 ± 3 counts were observed in the 14 channels where the ^{37}Ar should appear. The counter background for this period of time corresponded to 12 ± 4 counts. *Thus, there is no increase in counts from the sample over that expected from the background counting rate of the counter"* (emphasis added). The researchers calculated an upper limit of 0.5 captures per day, which corresponded to a limit of 3 SNU.

In an adjoining paper, Bahcall and collaborators published a new theoretical prediction for capture rate, one that was considerably reduced from previous estimates (Bahcall, Bahcall, and Shaviv 1968). It was 7.5 ± 3.0 SNU. They concluded, "The results of Davis, Harmer, and Hoffman are not in obvious conflict with the theory of stellar structure."

Edwin Salpeter summarized the situation (1968): "The present state of affairs is most frustrating for all concerned. The original theoretical estimate of about 12 counts per day would have been easily and accurately measurable and the theoretical revisions could as easily been up as down. They were down, however, and we have seen that a further factor of two in the theoretical estimate is quite possible. *Thus, at the present time, we neither have a positive identification of solar neutrinos nor the morbid satisfaction of predicting a scandal in stellar evolution"* (emphasis added).

Arguably, no serious discrepancy remained between experiment and theory because the experiment had set only an upper limit for solar neutrinos and because both the experimental result and the theoretical prediction had significant uncertainties. It was also possible that the apparatus could not detect solar neutrinos. That possibility did not remain for long. In 1970, a technical improvement to the analysis procedure allowed an actual measurement of the capture rate. The improvement was to use the rise time of the pulses produced by the decay electrons from ^{37}Ar. The pulses produced in the proportional counter by these electrons have a shorter rise time than those produced by background events. The experimenters could use both the pulse height, which is a measure of the electron energy, and the rise time to identify electrons from ^{37}Ar decay (Rowley, Cleveland, and Davis 1985).

The improvements in the experiment allowed a definite measurement of the neutrino capture rate rather than an upper limit. In 1970, starting with run 18 of the Homestake mine experiment, Davis and his collaborators reported experimental results that established the solar neutrino problem. During a period of 14 years, the result stabilized and the uncertainty became smaller. By 1984 the average of 61 separate experimental runs was 2.0 ± 0.3 SNU. At that time the best theoretical predictions were 6.6, 5.6, and 6.9 SNU, respectively, based on slightly different standard solar model (SSM) calculations.

Davis and his collaborators described the solar neutrino problem as "the discrepancy between the results of this experiment and the result predicted by solar model calculations using the best available input physics; i.e., by the standard solar model. The neutrino capture rate in the chlorine detector calculated using the standard solar model has changed with time as new data have become available. However, since 1969, in spite of great effort producing many new and improved measurements of nuclear reaction cross-sections, new opacity calculations etc. the capture rate has not changed in a major way" (Rowley, Cleveland, and Davis 1985).

Bahcall and Davis concurred (1989), "It is surprising to us, and perhaps more than a little disappointing, to realize that there has been very little qualitative change in either the observations or the standard theory since these papers appeared [1968], despite a dozen years of reexamination and continuous effort to improve details." The reason for their disappointment is shown clearly in figure 10.1, which displays the history of both experimental results and theoretical predictions as a function of time. There is a clear discrepancy between the results and the predictions.

More Solar Neutrino Experiments

Even before the Homestake mine chlorine experiment was completed, the difficulties of the method led to the suggestion by V. A. Kuz'min of another possible method of detecting solar neutrinos: namely, using a

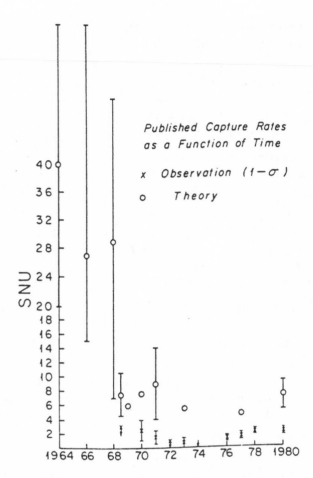

FIGURE 10.1. Published theoretical and experimental rates for the capture of solar neutrinos as a function of time. The experimental rates are clearly lower than those predicted theoretically. From Bahcall and Davis (1989).

gallium detector and detecting the solar neutrinos by the reaction ν_e + $^{71}Ga \rightarrow {}^{71}Ge + e^-$. Gallium had several advantages as a target. Perhaps most important was the fact that the threshold energy for the $^{71}Ga \rightarrow {}^{71}Ge$ reaction, 0.237 MeV, was low enough to detect virtually all of the solar neutrinos produced, whereas the chlorine detector was sensitive only to high-energy neutrinos from the decay of ^8B or ^7Be. Thus, a much higher neutrino capture rate would be expected in gallium than in chlorine. In addition, the cross section for the absorption of neutrinos by ^{71}Ga was expected to be quite large. The ^{71}Ge produced had a half-life of 11.4 days, which allowed a reasonable time for the extraction of the germanium from the gallium target without too much loss from radioactive decay. The energy of the electron emitted in germanium decay was approximately 12 keV (thousand electron volts), which was considerable higher than the approximately 2.5 keV of the electron in ^{37}Ar decay, allowing for the possibility of separating it more easily from background electrons.

Bahcall and his collaborators (1978) calculated the expected solar neutrino yield for a gallium detector for five different solar models; for neutrino oscillations, in which one type of neutrino transforms into another; and for possible neutrino decay. Four solar models gave expected neutrino capture rates of 80 ± 10 SNU for the gallium detector, far higher than the approximately 7 SNU predicted for the chlorine detector. The CNO model, which they regarded as very unlikely, gave an even higher rate, 487 SNU. Neutrino oscillations, the transformation of one type of neutrino into another, had a predicted rate of ≤31 SNU, and neutrino decay gave 0.

The detector would require approximately 50 tn of gallium to detect one capture per day. Bahcall and colleagues suggested two methods for separating the germanium produced from the gallium, calculated the backgrounds, and concluded, "A ^{71}Ga detector for low-energy solar neutrinos is feasible and desirable."

The first results from a gallium experiment appeared in 1991. The Soviet-American Gallium Experiment (SAGE) collaboration reported a measured neutrino capture rate of 20^{+15}_{-20} (statistical uncertainty [stat.]) ± 32 (systematic uncertainty [sys.]) SNU, with an upper limit of 79 SNU (Abazov et al. 1991). This was in contrast to a predicted value, at the time, of 132 SNU. The experimenters noted the previously reported negative results from the Homestake mine chlorine experiments and the recent results from the Kamiokonde II water–Cerenkov experiment that had "confirmed this deficit."

The experiment used a 30-tn liquid gallium detector. Because deep mines were not available in the Soviet Union, the detector was placed in a 4-km-long tunnel dug into Mount Andrychi in the Baksan Valley in the North Caucasus mountains. This reduced background produced by cosmic rays. After a typical exposure of 3–4 weeks, the germanium produced by neutrino capture was separated from the gallium

Not all of the data gathering went well, the researchers reported: "Results from measurements carried out in January, February, March, April, and July of 1990 are reported here. Earlier data taken during 1989 are not presented here due to the presence of radon and ^{68}Ge residual contaminations. The run during May of 1990 was unusable due to an instability in the electronics used and the run during June of 1990 was lost due to a vacuum accident." Experiments do not always run properly.

The scientists concluded, "The first measurements indicate that the flux may be less than that expected from p-p neutrinos alone. Thus, the solar neutrino problem may also apply to the low-energy p-p neutrinos, indicating the existence of new neutrino properties." These new neutrino properties included possible neutrino decay and neutrino oscillations.

The SAGE collaboration made improvements to the experimental apparatus, including, in mid-1991, an increase in its size to 57 tn, and continued taking data. This increase in size increased the capture rate and improved the statistical accuracy of the results. The values obtained for the neutrino capture rate were

1990	40^{+31}_{-38}	(stat.)	$^{+5}_{-7}$	(sys.)	SNU
1991	100^{+30}_{-26}	(stat.)	$^{+5}_{-7}$	(sys.)	SNU
1992	62^{+29}_{-27}	(stat.)	$^{+5}_{-7}$	(sys.)	SNU
Average	73^{+18}_{-16}	(stat.)	$^{+5}_{-7}$	(sys.)	SNU

(The theoretical prediction was 132 SNU.)

The experimenters noted that the change in the 1990 value from that previously reported (20^{+15}_{-20} (stat.) ± 32 (sys.)) "is due to a combination of revised counter efficiencies, the incorporation of the Earth-Sun distance correction, and the wider energy window used in the new standard analysis."

At the same time that the SAGE experiment was being performed, a second gallium experiment, GALLEX, was running in a tunnel beneath a mountain at Gran Sasso in Italy. The target contained 30.3 tn of gallium, of which 12 tn was ^{71}Ga. The detector was placed beneath a large amount of material to reduce background from cosmic rays. The significance of the experiment was clear to the researchers. They noted the low fluxes reported in the Homestake experiment and in the Kamiokonde detector, both of which were sensitive primarily to ^{8}B neutrinos.

> Both [experiments] have reported fluxes less than half the theoretical expectations. This discrepancy has become the so-called "solar neutrino problem." Various explanations have been suggested; among them are overestimates of the central temperature in the solar model calculations, or electron-neutrino modifications between Sun and detector, a manifestation of neutrino mass.
>
> The decision between these alternatives would be least ambiguous if a shortage of pp neutrinos were also observed. Given that the present Sun is producing fusion energy in equilibrium with its luminosity, a substantial reduction of pp neutrinos could only be due to some form of electron-neutrino disappearance. . . .
>
> In any case, the detection of pp neutrinos would be the first *experimental* proof of energy production by fusion inside the Sun. (Anselmann et al. 1992b)

The GALLEX group reported an initial result 83 ± 19 (stat.) ± 8 (sys.) SNU. This was in disagreement with the contemporaneous SAGE results of 20^{+15}_{-20} (stat.) ± 32 (sys.) SNU. "We have no explanation of the discrepancy between our results and those of the SAGE Collaboration," said the GALLEX researchers. That disagreement would disappear with the later SAGE results. The experimenters concluded that they had, in fact, observed neutrinos from proton-proton interactions for the first time. They also noted that, within two standard deviations, their result was in agreement with the full prediction. "However, to explain the GALLEX results plus those of ^{37}Cl and Kamiokonde in this way requires stretching the solar models and the data to their error limits; yet the possibility remains open." They also stated that their result was consistent with neutrino oscillations.

The simultaneous agreement of the later SAGE results and those of GALLEX strengthened their credibility: "The measurements of GALLEX during 1991–1993, made with a different form of Ga target (aqueous GaCl$_3$ solution), observes 79 ± 10 ± 6 SNU. The good agreement of the measurements of *two independent Ga experiments with different forms of the target material is quite important and gratifying"* (emphasis added). Obtaining the same result with different experiments is more credible than merely getting the same result from repeating the same experiment.

In April 1992 the gallium solution was transferred from the original detector to a different tank, and data taking continued (GALLEX II). In 1994 the experimenters concluded that the GALLEX result, 79 ± 12 SNU, was significantly lower than the predictions of the SSM, but that taken by itself it did not provide enough information to establish whether the discrepancy was due to an error in the SSM or to some other neutrino properties (Anselmann et al. 1994).

The Kamiokonde II Experiment

The fourth solar neutrino experiment was the Kamiokonde water–Cerenkov counter experiment. Unlike the Homestake mine chlorine experiment and the gallium experiments, which detected the presence of solar neutrino interactions only long after they occurred, the Kamiokonde apparatus detected neutrino interactions in real time, when they occurred. It did so by detecting the Cerenkov radiation emitted by the electrons produced in neutrino-electron collisions. (Cerenkov radiation is produced when a charged particle travels faster than light in a medium.) Because the threshold energy for detection was 9.3 MeV (later 7.5 MeV), the Kamiokonde detector was sensitive only to neutrinos from ^{8}B decay. The detector provided information on the neutrino arrival time, its direction, and the energy spectrum of the electrons produced.

The results for the first 1,040 days of data taking are shown in figure 10.2 (Hirata et al. 1991). There is a clear peak near $\cos\theta \approx 1$ ($\theta \approx 0°$), indicating that the events were caused by neutrinos from the sun. The experimenters also checked that the signal was a "real" signal. "The analysis was repeated with the detector location assigned to other, incorrect, latitudes and longitudes. The signal peaks only when the true

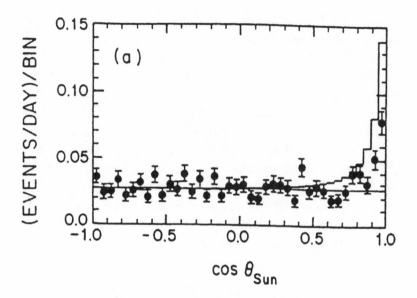

FIGURE 10.2. Plot of the cosine of the angle between the electron direction observed in the Kamiokonde detector and a radius vector from the sun to Kamiokonde showing the signal from the sun plus an isotropic background. The dots are the experimental results, and the solid curve is the theoretical prediction. The result is much lower than predicted. BIN indicates the number of events for intervals of $\cos\theta_{sun}$, for example, between approximately 0.02 and 0.04. From Hirata et al. (1991).

Kamiokonde coordinates are assigned," the researchers noted. The data, however, were well below what was expected theoretically (solid curve): data/SSM = 0.46 ± 0.05 (stat.) ± 0.06 (sys.) was in clear disagreement with the predictions of the SSM.

This result was further checked by an independent analysis procedure: "Two independent analyses were performed on the same data. Each analysis obtained the final sample using totally independent programs for the event reconstruction and applied different cuts [selection criteria]."

A question was also raised about whether the Kamiokonde results were consistent with those obtained in the Homestake mine chlorine experiment. These two experiments were sensitive to similar groups of neutrinos: the Homestake experiment was sensitive to neutrinos from

TABLE 10.1
Summary of early solar neutrino experiments

	SAGE + GALLEX	Chlorine	Kamiokonde
Target material	^{71}Ga	^{37}Cl	H_2O
Reaction	$v_e + {}^{71}Ga \rightarrow {}^{71}Ge + e^-$	$v_e + {}^{37}Cl \rightarrow {}^{37}Ar + e^-$	$v_e + e^- \rightarrow v_e + e^-$
Detection method	Radiochemical	Radiochemical	Cerenkov
Detection threshold	0.234 MeV	0.814 MeV	7.0 MeV
Neutrinos detected	All	^{7}Be and ^{8}B	^{8}B
Predicted rate	132 ± 7 SNU	9 ± 1 SNU	5.7 ± 0.8 flux units[a]
Observed rate	74 ± 8 SNU	2.5 ± 0.2 SNU	2.9 ± 0.4 flux units

Note: 1 SNU (solar neutrino unit) = 10^{-36} captures per target atom per second.
[a]In units of 10^6 neutrinos per square centimeter per second.

both ^{7}Be and ^{8}B, whereas the Kamiokonde experiment detected only those from ^{8}B. The experimenters stated, "No significant disagreement exits between the two data sets." They further noted, "Thus, it is difficult to explain the results of both the Kamiokonde II and the ^{37}Cl detectors (assuming both are correct) by manipulating the solar model. This in turn suggests that some as yet undetected intrinsic property of neutrinos might be playing a role in the solar neutrino deficit." Neutrino decay or oscillations might be the culprit.

The entire evidential situation for the SSM was even worse than just the Kamiokonde and Homestake results. The results from both gallium experiments were also in disagreement with the SSM. Thus, four experiments disagreed with the SSM (table 10.1). Although there were slight differences among various solar model calculations, none were anywhere close to being large enough to solve the solar neutrino problem. Bahcall and R. K. Ulrich, for example, constructed 1,000 different SSMs by randomly varying, within their given uncertainties, the parameters used in the SSM (1988). The results of their calculations and neutrino fluxes required to explain the Homestake mine result could not be constructed within plausible uncertainties in the SSM. Something else was going on.

THE MISSING SOLAR NEUTRINOS   159

Neutrino Oscillations

Theory

The fact that the observed flux of solar neutrinos was far less than that predicted by the SSM could not be explained by modifications of that model. Two alternative explanations were offered. The first was neutrino decay. If neutrinos have a finite lifetime (less than approximately 8 min, the time of travel between the sun and the earth, because otherwise a significant number would not decay before the neutrinos reached the earth), then the solar neutrino deficit could be explained by the fact that the neutrinos decayed before reaching the earth. This suggestion was rejected when neutrinos from the supernova SN 1987A were observed on earth. If the neutrino lifetime is short enough to explain the lack of solar neutrinos, then no neutrinos should be observed from the far more distant supernova. "The idea that neutrino decay into some sterile form [a noninteracting neutrino] might provide an explanation of the solar neutrino problem died in its most straightforward form along with the supernova SN 1987A, since the observation of (anti)neutrinos from that stellar explosion clearly requires survival times much longer than the Sun to Earth" (Anselmann et al. 1992a).

The second suggested alternative was neutrino oscillations: the idea that one type of neutrino can transform into a second type. For example, the electron neutrino might transform into a muon neutrino, and vice versa. This could explain the discrepancy between theory and experiment because the initial solar neutrino experiments were sensitive only to electron neutrinos, and some of them would be lost if, during the travel from the sun to the earth, electron neutrinos transform into muon neutrinos that could not be detected by those early experiments.

In 1958 only the existence of the electron neutrino was known, and Pontecorvo (1958) speculated that neutrino-antineutrino oscillations would be possible. Pontecorvo returned to this issue in 1967. By this time the muon neutrino had been discovered, and he speculated that oscillations of the type ($\nu \leftrightarrow \bar{\nu}, \nu_\mu \leftrightarrow \nu_e$) would become possible for neutrino beams (Pontecorvo 1968). At the time of this paper, the results of the Homestake mine chlorine experiment were not yet known.

When they became known in 1968, Pontecorvo and V. Gribov considered the possibility that the solar neutrino deficit might be explained by neutrino oscillations. "It is shown that lepton nonconservation might lead to a decrease in the number of detectable solar neutrinos at the earth surface, because of $v_e \leftrightarrow v_\mu$ oscillations" (Gribov and Pontecorvo 1969).

The work of Gribov and Pontecorvo was extended by Lincoln Wolfenstein (1978), who specified conditions under which neutrino oscillations could occur. One prediction of Wolfenstein's calculation was that neutrino oscillations would cause a reduction in the number of solar neutrinos, although not a large enough reduction to solve the solar neutrino problem. S. P. Mikheyev and A. Y. Smirnov further extended Wolfenstein's calculation. They found that "matter can enhance neutrino oscillations" and that they could reduce the predicted neutrino fluxes for both the chlorine and gallium experiments by a factor of three (1985, 1986). It seemed that such resonant oscillations, later known as the MSW effect (for Mikheyev, Smirnov, and Wolfenstein), might be able to solve the solar neutrino problem. The next step was to see if such oscillations would be observed.

Experimental Tests

The first attempts to fix the parameters of possible neutrino oscillation came from analyses of the solar neutrino experiments themselves. As the GALLEX collaborators remarked, "If we accept the solar model as given, we can analyze all discrepancies between solar model calculations and the results of the solar neutrino experiments in terms of alteration of the properties of solar neutrinos during their passage from the solar core to the detector" (Anselmann et al. 1992a).

Several groups analyzed the results using only the deficiency in the flux of solar neutrinos. The Kamiokonde researchers used not only that deficiency but also the observed electron energy spectrum of the recoil electrons produced in neutrino-electron interactions in their detector (Hirata et al. 1988). That energy spectrum depended in turn on the initial neutrino energy spectrum, and that could also be used to fix the oscillation parameters. Using the electron energy spectrum gave increased sensitivity. Further analysis was provided by the GALLEX experimenters. They combined the results, as of 1992, of the Homestake chlorine experiment, the Kamiokonde experiment, and their own gallium experiment and obtained a good fit to the oscillation parameters (Anselmann et al. 1992a). Neutrino oscillations had become the sole remaining plausible explanation of the deficiency in the number of solar neutrinos, an example of the Sherlock Holmes strategy. The physics community, however, wanted evidence in favor of neutrino oscillations that was independent of the SSM.

Initially, the most important evidence on neutrino oscillations came from the experiments on atmospheric neutrinos. Although other experiments contributed to the discussion, the series of experiments performed with the Kamiokonde detectors was key to the outcome and initially provided the most persuasive evidence in favor of neutrino oscillations. In the end, experiments done at the Sudbury Neutrino Observatory (SNO) provided the definitive results that solved the solar neutrino problem.

The Kamiokonde researchers described the basic physics underlying their experiment: "Primary cosmic rays striking the atmosphere produce pions and kaons which subsequently decay into muons and muon-neutrinos, and much less abundantly, electrons and electron-neutrinos. The muons further decay into electron-neutrinos and muon-neutrinos. As a consequence, it is expected that there are roughly two muon-neutrinos for each electron neutrino and that the shape of the energy spectrum is similar to that of the pions and muons. [The main reactions are $\pi \to \mu + \nu_\mu$ followed by $\mu \to e + \nu_\mu + \nu_e$.] *Since atmospheric neutrinos must traverse the earth almost freely, they enter the detector from every direction*" (Hirata et al. 1988; emphasis added). An asymmetry in the number of neutrinos with respect to their direction relative to the earth would become crucial evidence in favor of neutrino oscillations.

The neutrinos were to be detected by the number of electrons and muons produced by their interaction with matter. More specifically, both the electrons and muons produced by neutrino interactions with the protons in the water would produce Cerenkov radiation, which would be detected by the phototubes in the Kamiokonde detector. The experimenters assumed that lepton family number would be conserved, that is, that electron neutrinos (antineutrinos) would produce electrons (positrons) and that muons and their neutrinos would behave similarly. Two of the crucial issues in determining a result were the separation of electrons from muons and the calculation of the expected number of electrons and muons given the estimated neutrino fluxes. Considerable care was devoted to these issues.

The experimenters compared the observed number of electron-like and muon-like events to the estimates provided by the Monte Carlo simulation of the experiment. The number of observed electron-like events was (105 ± 11) percent of the number predicted by the Monte Carlo simulation. For muon-like events, the value was (59 ± 7) percent. There was an additional observation that supported the low number of muon-like events. This was the observation of muon-like events followed by a muon-decay electron. These electrons should, of course, appear only for muon events. The observed number of muon-like events accompanied by a muon-decay electron was 52, whereas 103.8 were expected. Muon events, and therefore muon neutrinos, seemed to be disappearing.

The experimenters questioned whether the Monte Carlo simulation, on which the number of events expected was based, was correct. They also searched for other possible systematic effects that might preferentially reduce the number of muon-like events, but noted, "We have as yet found no effect that reproduces the deficiency of muon-like events relative to the total of electron-like events."

The experimenters concluded, rather tentatively, "Some as-yet-unaccounted-for physics might be necessary to explain the result. Neutrino oscillations between muon-neutrino and v_x or between electron-neutrinos and muon-neutrinos might be one of the possibilities that could explain the data."

In 1992, the Kamiokonde researchers reported a new result based on additional data (Hirata et al. 1992). They remarked that their earlier result had not taken into account the effect of muon polarization in pion decay in their calculation of the neutrino flux, although the effect was not large enough to explain their observed deficit of muon neutrinos. Their new result for the ratio of muon events to electron events,

$R(\mu/e)$, was $0.67^{+0.07}_{-0.06}$ (stat.) $\pm$ 0.05 (sys.). They once again emphasized the importance of particle identification: "Among the possible
systematic errors, the most critical is the error in the μ/e identification
since it affects the numerator and denominator in $R(\mu/e)$ in opposite
directions." They stressed their continuing checks on this with the use
of stopping muons and the number of muon decay events. In particular, they noted that their result for $R(\mu/e)$ obtained from muon-decay
events was 0.61 $\pm$ 0.07, "in excellent agreement with the value obtained
from particle identification." Their new result was also consistent with
their previous result and with the most recent result from IMB-3 (another experiment) of $R(\mu/e)$ = 0.61 $\pm$ 0.07.

The Kamiokonde researchers now considered neutrino oscillations
more seriously as a possible explanation for the muon neutrino deficit.
They concluded, "Neutrino oscillations, with the totality of atmos-
pheric and solar neutrino data favoring the channel $\nu_\mu \leftrightarrow \nu_\tau$, might
account for the observations" (Hirata et al. 1992).

The Kamiokonde experimenters continued their efforts, upgrading
the detector to Kamiokonde III (Fukuda et al. 1994). They were then able
to detect events in the multi-GeV (billion electron volts) region. (Their
previous experiments had detected neutrinos with energy less than 1.33
GeV.) Their result for the multi-GeV events was $((\mu/e)_{data}/(\mu/e)_{MC})$ =
$0.57^{+0.08}_{-0.07}$ (stat.) $\pm$ 0.07 (sys.) and $0.59^{+0.08}_{-0.07}$ (stat.) $\pm$ 0.07 (sys.),
respectively, for two different neutrino flux calculations. These results
were not only mutually consistent, but also consistent with the collaborators' previous low-energy result, $0.60^{+0.06}_{-0.05}$ (stat.) $\pm$ 0.07 (sys.)
and $0.61^{+0.06}_{-0.05}$ (stat.) $\pm$ 0.07 (sys.) for the two calculated fluxes. In
contrast to the symmetric low-energy data, the multi-GeV events
showed a marked up-down asymmetry. A good fit was obtained for
neutrino oscillations, with either $\nu_\mu \leftrightarrow \nu_e$ or $\nu_\mu \leftrightarrow \nu_\tau$. "The data were
analyzed assuming neutrino oscillations and yielded allowed regions
of the oscillation parameters for both $\nu_\mu \leftrightarrow \nu_e$ or $\nu_\mu \leftrightarrow \nu_\tau$ channels,
consistent with those obtained from our sub-GeV data," they wrote.
The $\nu_\mu \leftrightarrow \nu_e$ oscillations remained a possibility.

There were still questions about whether there were background
processes that might produce the observed deficit in R. One such possible source of background, which might decrease the observed ν_μ/ν_e
rate, was neutrons produced by cosmic ray muons. The neutrons would

then produce neutral pions, which would then decay into two γ-rays. The γ-rays would then produce electromagnetic showers in the detector, which would simulate e-like events. The Kamiokonde researchers investigated this question by examining the vertex position distribution of neutral-pion-like events. They concluded, "No evidence for the background contamination was observed" (Fukuda et al. 1994).

Super Kamiokonde, a water Cerenkov detector approximately 10 times larger than Kamiokonde, began operations, and the first results on atmospheric neutrinos appeared in 1998 (Fukuda et al. 1998). The larger detector acquired data at a much faster rate and provided better evidence in support of neutrino oscillations. The particle-identification method was tested with electrons from muon decay and stopping cosmic ray muons. "The μ-like and e-like events used in this analysis are clearly separated," they noted. In addition, "the data were analyzed independently by two groups, making the possibility of significant biases in data selection or event reconstruction algorithms remote."

Neutrino oscillations were clearly shown in the graph of the up-down asymmetry, $(U - D)/(U + D)$ as a function of momentum (figure 10.3). U is the number of upward-going events (these neutrinos passed through the earth), and D is the number of downward-going events. The electron-like events were consistent with expectations, "while the μ-like asymmetry is consistent with zero, but significantly deviates from zero at higher momentum," they wrote. The neutrino oscillations took place in the extra path length traveled by the upward-going neutrinos. The average path lengths were 15 km for downward-going neutrinos, as compared with approximately 13,000 km for upward going. Further analysis of the asymmetry argued against $v_\mu \leftrightarrow v_e$ oscillations.

The experimenters concluded, "Both the zenith angle distribution of μ-like events and the value of R observed in this experiment significantly differ from the best predictions in the absence of neutrino oscillations. . . . We conclude that the present data give evidence for neutrino oscillations." The result favored $v_\mu \leftrightarrow v_\tau$ oscillations. The Kamiokonde results have been confirmed by virtually all of the experiments on atmospheric neutrinos.

Very persuasive evidence came from Kamiokonde, and from other experiments, on atmospheric neutrinos for oscillations of the muon neutrino into the τ-neutrino. Unfortunately, that did not solve the solar

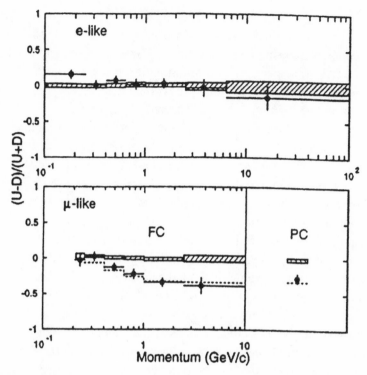

FIGURE 10.3. The (up − down)/(up + down) asymmetry as a function of momentum for two samples of e-like and μ-like events. The Monte Carlo expectations without neutrino oscillations are shown in the hatched region. The dashed line for μ-like events is the expectation for $\nu_\mu \leftrightarrow \nu_\tau$ oscillations. FC, fully contained events; PC, partially contained events. From Fukuda et al. (1998).

neutrino problem, which is a deficit of electron neutrinos. The experimental results on $\nu_\mu \leftrightarrow \nu_e$ oscillations, which could solve the problem, were, at the time, generally negative but not totally conclusive. Thus, the solar neutrino problem remained unsolved.

The Solution

The situation changed dramatically in the summer of 2001 when striking evidence for electron neutrino oscillations was provided by the SNO (Ahmad et al. 2001). The SNO was an imaging water Cerenkov detector, similar to Kamiokonde, located underground in the Creighton mine near Sudbury, Ontario, Canada. The detector contained ultrapure heavy water (D_2O rather than H_2O; D is deuterium, heavy hydrogen) and was sensitive to the following reactions.

$$\nu_e + d \rightarrow p + p + e^- \quad \text{(charged current [CC])}$$
$$\nu_x + d \rightarrow n + p + \nu_x \quad \text{(neutral current [NC])}$$
$$\nu_x + e^- \rightarrow \nu_x + e^- \quad \text{(elastic scattering [ES])}$$

The earlier experiments were not sensitive to either the CC or NC reactions because the detectors contained water, not heavy water. "The charged current (CC) reaction is sensitive exclusively to electron-type neutrinos, while the neutral current (NC) is sensitive to all active neutrino flavors ($x = e, \mu, \tau$). The elastic scattering (ES) reaction is sensitive to all flavors as well, but with reduced sensitivity to ν_μ and ν_τ," the paper said. The experimenters also remarked that if electron neutrinos from the sun change into other types of neutrino, then the flux of neutrinos calculated from the CC reaction, $\phi^{CC}(\nu_e)$, would be less than that calculated from the ES reaction, $\phi^{ES}(\nu_x)$.

The initial SNO results were on the CC and ES reactions. Their ES result was consistent with that obtained earlier by Super Kamiokonde. The ^8B neutrino flux measured by the CC reaction $\nu_e + d \rightarrow p + p + e^-$, a reaction sensitive only to electron neutrinos, was $\phi^{CC}(\nu_e) = 1.75 \pm 0.07$ (stat.) $^{+0.12}_{-0.11}$ (sys.) ± 0.05 (theoretical) $\times 10^6$ cm^2 s^{-1}. (In addition to the statistical and systematic experimental uncertainties, the experimenters included an estimate of the uncertainty in theoretical calculations.) The value of the ^8B neutrino flux measured by the Super Kamiokonde collaboration for the ES reaction $\nu_x + e^- \rightarrow \nu_x + e^-$, which is sensitive to all three types of neutrinos, is $\phi^{ES}(\nu_x) = 2.32 \pm 0.03$ (stat.) $^{+0.08}_{-0.07}$ (sys.) $\times 10^6$ cm^2 s^{-1}. The difference between the two measured fluxes indicates the presence of other types of neutrino in the solar flux observed at the earth and is strong evidence for electron neutrino oscillations. "The measurement of $\phi^{CC}(\nu_e)$, however, is significantly smaller [than $\phi^{ES}(\nu_x)$], and is therefore inconsistent with the null hypothesis that all observed solar neutrinos are ν_e. . . . These data are therefore evidence of a nonelectron active flavor component [other types of neutrino] in the solar neutrino flux," the researchers concluded.

In addition, the SNO collaboration determined that the total flux of active ^{8}B neutrinos was $5.44 \pm 0.99 \times 10^6$ cm^2 s^{-1}, in "excellent agreement" with the predictions of the SSM of 5.05×10^6 cm^2 s^{-1}.

Additional evidence for neutrino oscillations was reported in the summer of 2002. The SNO collaboration reported results for all three reactions: CC, ES, and NC (Ahmad et al. 2002). The last two are sensitive to all three types of neutrino. "Using the neutral current (NC), elastic scattering, and charged current reactions and assuming the standard ^{8}B shape, the ν_e component of the ^{8}B solar flux is $\phi_e = 1.76 \pm 0.05$ (statistical) ± 0.09 (systematic) $\times 10^6$ cm^{-2}s^{-1}. . . . The non-ν_e component is $\phi_{\mu\tau} = 3.41 \, ^{+0.45}_{-0.48}$ (stat) $^{+0.45}_{-0.48}$ (syst) 10^6 cm^{-2}s^{-1}, 5.3σ [standard deviations] greater than zero, providing strong evidence for solar ν_e flavor transformation. The total flux measured with the NC reaction is $\phi_{NC} = 5.09^{+0.44}_{-0.43}$ (stat) $^{+0.46}_{-0.43}$ (syst) 10^6 cm^{-2}s^{-1}, consistent with solar models," the experimenters wrote.

The solar neutrino problem had been solved after more than 30 years. The deficit was due to neutrino oscillations.

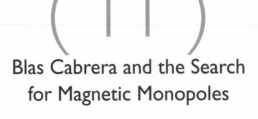

Blas Cabrera and the Search
for Magnetic Monopoles

One of the interesting problems concerning experimental results is what happens when an experiment gives a null result, when the phenomenon expected is not observed. It could happen because the experimental apparatus and the associated analysis procedures cannot detect or measure the phenomenon in question or because the phenomenon is not present. This is a real problem in physics investigations. In the Michelson-Morley experiment, one of the most famous experiments in modern physics, Albert Michelson and Edward Morley expected to detect a shift in a pattern of light caused by the motion of the earth relative to the ether (the proposed medium though which electromagnetic waves traveled). They observed no such shift. Possibly it was because the apparatus was faulty or because the earth's velocity relative to the ether was zero. Other explanations were later offered for this null result. They included the ether-drag hypothesis, that the earth dragged a layer of the ether along with it; the Lorentz-Fitzgerald contraction, that an object shrank in the direction of its motion relative to the ether; and the ballistic theory, that the velocity of light was constant relative to the source of the light. Eventually, all of these alternatives were rejected on the basis of experimental evidence, and it was accepted that the earth had zero velocity with respect to the ether. Later, Albert Einstein's special theory of relativity dispensed with the need for the ether and explained the result of the Michelson-Morley experiment.

A related problem occurs when an effect is initially observed but later—and presumably better—experiments do not replicate the observation. Then experimenters have to argue that these later experiments are indeed better and that their results are more reliable.

The history of Blas Cabrera's searches for magnetic monopoles (single, north or south magnetic poles) is fascinating because in his first experiment, Cabrera observed a signal that not only was consistent with the existence of a magnetic monopole, but was precisely the size that theory predicted. Despite this observation, Cabrera made no discovery claim. That was because he could not rule out all alternative causes for his observed signal. Cabrera made several improvements to his experimental apparatus and continued the search. None of the later searches found evidence for magnetic monopoles. Cabrera and his colleagues used several arguments to demonstrate that these later experiments were in fact better and that they would have observed magnetic monopoles had they been present. They then concluded, quite reasonably, that magnetic monopoles had not been observed.

Perhaps the most typical strategy used by experimenters to demonstrate that their apparatus would have observed the phenomenon in question is the use of a surrogate signal. Producing an adequate surrogate can involve considerable ingenuity by physicists. Some commentators on science have questioned whether the adequacy of such surrogate signals is examined in sufficient detail (see, for example, Collins 1985, 1994). I have argued elsewhere in detail that scientists do consider the question of the adequacy of a surrogate signal quite carefully and do present arguments for their validity (Franklin 1997b). In the magnetic monopole investigation, the creation and use of a surrogate signal not only helped to establish the ability of the apparatus to detect the phenomena, but also helped to cast doubt on the initial observation.

Are There Magnetic Monopoles?

One interesting fact about electromagnetism is that single electric charges, positive and negative, exist, whereas single magnetic charges —magnetic monopoles—do not. All known magnetic fields have two poles, north and south.

In 1931 Paul Dirac, winner of the Nobel Prize for his work on relativistic quantum theory, began a theoretical investigation that led to interesting conclusions about magnetic monopoles, if they existed. Dirac's original intent was to try to provide a reason for the existence of the smallest unit of electric charge, e, the charge of the electron. (All known charges are integral multiples of e.) His paper, he wrote, "will be concerned essentially, not with electrons and protons, but with the reason for the existence of the smallest electric charge. This smallest charge is known to exist experimentally and to have the value e given approximately by $hc/2\pi e^2 = 137$ [h is Planck's constant and c is the speed of light]. The theory of this paper, while it looks at first as though it will give a theoretical value for e, is found when worked out to give a connection between the smallest electric charge and the smallest magnetic pole. It shows, in fact, a symmetry between electricity and magnetism quite foreign to current views."

Dirac had found something surprising in his theory: that the fundamental unit of electric charge was related to the fundamental unit of magnetic pole. "The theory leads to a connection, namely equation (9), between the quantum of magnetic pole and the electric charge." Dirac's equation (9) was $hc/2\pi eg = 2$, where g was the strength of the magnetic pole. Dirac further noted, "This means that the attractive force between two one-quantum poles of opposite sign is $(137/2)^2 = 4692\frac{1}{4}$ times that between electron and proton. This very large force may perhaps account for why poles of opposite sign have never yet been observed." The force was too large to easily separate the poles.

In later work Dirac presented a more general theory of the interaction of charged particles and magnetic poles. "If one supposes that a particle with a single magnetic pole can exist and that it interacts with charged particles, the laws of quantum mechanics lead to the requirement that the electric charge be quantized—all charges must be integral multiples of a unit charge e connected with the pole strength g by the formula $eg = hc/4\pi$. Since electric charges are known to be quantized and no reason for this has yet been proposed apart from the existence of magnetic poles, we have a reason for taking magnetic poles seriously. The fact that they have not yet been observed may be ascribed to the large value of the quantum of pole" (1948).

This theoretical work formed the background to searches for magnetic monopoles and provided an enabling theory for the experiments by giving an estimate of the strength of the magnetic pole and of the size of the effects that might be observed.

The Saint Valentine's Day Event

One of the interesting searches for magnetic monopoles was conducted by Blas Cabrera and his collaborators in the 1980s and 1990s. In the first experimental run, an event consistent with a monopole was found, but all subsequent searches by the group, with improved apparatus, found no similar event. Was the first event an example of very rare magnetic monopoles, or was it an artifact of the experiment?

Cabrera's method of searching for magnetic monopoles was conceptually straightforward although technically difficult, particularly for large-area detectors. He used a loop of superconducting wire connected to the superconducting input coil of a SQUID (superconducting quantum interference device) magnetometer. A magnetic monopole passing through such a loop of superconducting wire will produce a change in magnetic flux through the loop of $4\pi g = hc/e$, where g is the magnetic charge of the monopole, h is Planck's constant, c is the speed of light, and e is the charge of the electron. This is twice the flux quantum of superconductivity $\Phi_0 = hc/2e$. (Magnetic flux in a superconducting loop is quantized. It comes in integral multiples of a fundamental unit.) "Such a detector measures the moving particle's magnetic charge regardless of its velocity, mass, electric charge, or magnetic dipole moment. . . . In the general case, any trajectory of a magnetic charge g which passes through the ring will result in a flux-quanta change of 2, while one that misses the ring will produce no flux change," Cabrera wrote (1982).

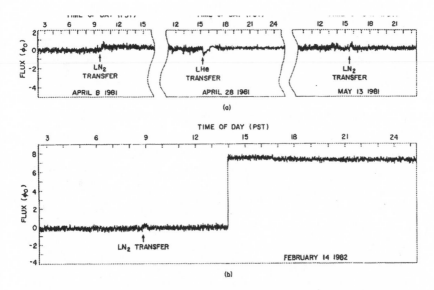

FIGURE 11.1. The Saint Valentine's Day event. Data records showing a, typical stability and b, the candidate monopole event. From Cabrera (1982).

Cabrera constructed a 20-cm^2 superconducting loop and took data for 151 days. The loop contained four turns, so, he noted, "The passage of a single Dirac charge through the loop would result in an 8-Φ_0 change in flux through the superconducting circuit, comprised of the detection loop and the SQUID input coil (a factor of two from $4\pi g = 2\ \Phi_0$ and a factor of four from the number of turns in the pickup loop)." The detector was calibrated in three different and independent ways, and the calibrations agreed within the stated uncertainties. Figure 11.1 shows several intervals of data recording. There were typical small disturbances in the trace due to the daily liquid-nitrogen transfer and weekly liquid-helium transfers. The transfers were necessary because the apparatus had to be at a temperature near absolute zero to operate properly. The disturbances were far smaller than that observed for the possible monopole event. "A single large event was recorded [figure 11.1b]. It is consistent with the passage of a single Dirac charge [8 Φ_0] within a combined uncertainty of ±5%. . . . It is the largest event of any kind in the record," Cabrera wrote. This event was recorded on February 14, 1982, when the laboratory was unoccupied, allowing for the possibility of a transient apparatus malfunction or even a human intervention.

Other flux changes were recorded. A total of 27 events exceeded a threshold of 0.2 Φ_o, after exclusion of known disturbances such as liquid-helium and liquid-nitrogen transfers. No other event was within a factor of four of the signal from the single large event, or that expected for a Dirac monopole.

Cabrera devoted considerable effort to searching for possible spurious detector responses that might have caused the possible monopole signal. All could be eliminated. However, one possible alternative explanation of the signal could not be conclusively rejected. "Mechanically induced offsets have been intentionally generated and are probably caused by shifts of the four-turn loop-wire geometry which produce inductance changes. *Sharp raps with a screw driver handle against the detector assembly cause such offsets. On two occasions out of 25 attempts these have exceeded 6Φ_0 (75% of the shift expected from one Dirac charge)*; however, drifts in the level were seen during the next hour" (emphasis added).

Cabrera did not think that this was a likely cause of the observed signal, but he did not feel that he could completely eliminate it as a possibility: "A spontaneous and large external mechanical impulse is not seen as a possible cause for the event; however, the evidence presented by this event does not preclude the possibility of a spontaneous internal stress release mechanism." He was, however, able to set an upper limit for the monopole flux.

In a later comment, Cabrera remarked, "It was a striking event, because it was exactly the right step size [for a Dirac monopole], but I was not convinced because of the other possible although improbable mechanism" (quoted in Staley 1999). Cabrera made no discovery claim for a magnetic monopole, but it remained a possibility.

Later Experiments

Cabrera and his collaborators continued the search for magnetic monopoles with an improved and larger experimental apparatus. The detector consisted of three superconducting loops of two turns each. The area of the new apparatus was 476 cm², with a loop area of 70.5 cm² and a near-miss area of 405 cm². (Monopoles striking the near-miss area would also register in the detector.) It was more than 20 times larger than Cabrera's original detector. The apparatus also included a calibration coil. When current was passed through the coil, it created a magnetic field that was detected by the superconducting loops. The experimenters found that a current of 53.2 nA (nanoamperes) in the calibration coil induced a supercurrent change of $4\,\Phi_0$, in all three loops. This was the signal expected for a Dirac monopole.

One of the most significant improvements to the experiment was the use of coincidence signals between the loops. Operation of the original noncoincidence single-loop detector demonstrated the need for discrimination against spurious events. Along with monitoring other known causes of spurious signals, the most reliable technique is to use coincidence detection, having two or more uncoupled detectors that will respond in coincidence to a monopole event but not to a spurious event. The coincidence requirement made the probability of detecting simultaneous spurious signals in two loops very small. Unlike the original experiment, in which some of the possible causes of spurious signals were checked only after the data were taken, in this experiment the experimenters monitored such causes as they went along.

Once again the researchers checked on the possibility that a mechanical effect could cause a spurious signal. They found a significant improvement. The signals produced were considerably smaller than they had been in the original experiment, and they were detected by the monitoring instruments. "Superconducting offsets can be generated by tapping on the Dewar [the vacuum chamber] with a mallet, but these signals also show up on the accelerometer data," they noted (Gardner et al. 1991). The experimenters found that "no events satisfied the double-coincidence requirement . . . [and] no large or spurious signals were seen, casting no light on the origin of the previously reported candidate. However, these data lower that previous flux limit by a factor of 38, increasing the probability of a spurious cause for that event" (Cabrera et al. 1983). The experimenters remarked that they planned to continue operating the detector for at least a year and were also designing a larger detector.

Cabrera and his group continued their search for magnetic monopoles with a further-improved eight-loop detector (Huber et al. 1990, 1991). They noted that a number of groups, including their own, had been searching for monopoles for several years without observing any convincing candidates.

Cabrera and his collaborators reported on the first 547 days of operation of their new eight-loop detector, in which each of the loops was located on the face of an octagonal prism. The total usable area of the detector was 1.1 m², "the largest superconducting detector to date" (Huber et al. 1990). This detector was 500 times larger than the original detector, thus increasing the probability of detecting a monopole. The experimenters also required a coincidence signal from two of the loops: "A feature of this geometry is that a monopole can induce a signal in at most two loops and, for most of the cross section, no fewer than two loops. In contrast, offsets in more than two loops must be the result of electrical or mechanical disturbances and are rejected as monopole candidates" (Huber et al. 1990). As is usually the case in real experiments, not everything went as planned. The original area of the detector was 1.5 m². "Upon cooling the detector, the [super]conducting NbTi ribbon cracked and opened two pickup-loop circuits, causing those loops sections to be unresponsive to flux changes" (Huber et al. 1990). This reduced their active sensing area to 1.3 m². It was later reduced to 1.1 m².

Once again the experimenters installed checks to guard against spurious signals. These were even more extensive than those used in the three-loop experiment.

> We installed additional instrumentation to monitor parameters known to affect detector operation. This instrumentation includes a strain gauge attached to the exterior of the superconducting lead shield (to detect mechanical motion), a pressure transducer (to monitor the helium pressure above the liquid in the Dewar), and a power line monitor (to detect six different fault conditions).... An ultrasonic motion detector [which would detect the presence of an intruder] monitors laboratory activity. When we perform activities known to disturb detector stability, we set a "veto" switch to prevent generating large numbers of useless computer events and to aid in calculating our live time. (Huber et al. 1990)

Other changes were introduced to reduce the number of spurious signals from known causes. In the earlier versions of the experiment, transfers of liquid helium and liquid nitrogen caused offsets. In this experiment, "A closed-cycle helium liquefier connected to our Dewar eliminates helium transfers and maintains a constant liquid-helium level, so the operation can be extremely stable. Gas-cooled radiation shields eliminate the need for liquid nitrogen."

A new calibration system was installed that more closely approximated the signal expected for a magnetic monopole. This consisted of long, small-diameter toroidally wound coils. These coils had two poles, but they were separated by a substantial distance so that a single pole could be inserted in the superconducting loops. Each calibration coil coupled to two adjacent loops simultaneously. The researchers explained, "A current of 0.19 μA through the coil . . . produces a flux equivalent to that of a Dirac monopole" (Huber et al. 1991; figure 11.2). The calibration signal was similar to that observed in the early monopole candidate event (figure 11.1). The safeguards against possible spurious signals were quite effective. At 9:41 on July 17, 1987, for example, a single-loop event was observed with a clear offset signal. Although it was not a monopole candidate because it was detected in only one loop of the detector, simultaneous signals were also detected in a strain gauge, a magnetometer, and a motion detector. These signals would also have been grounds for rejecting the event. It was spurious.

The collaborators detected 43 single-channel events that did not correlate with any disturbances in the monitors. On the basis of that frequency of occurrence, they calculated that accidental double coincidences would be detected approximately once in 800 years.

The incidence of double-coincident offsets is much higher than this estimate. Four have already been observed . . . however, the magnitudes are inconsistent with a Dirac charge, and such effects always occur in adjacent detectors. Since adjacent-pane events contribute only ~0.152 of the total sensing area it is extremely unlikely that we would observe four such events without observing events of any other type. The probability is approximately $(0.152)^4$, or 0.0005. A more likely cause is mutual rf [radiofrequency] interference between SQUID's coupled through adjacent pickup coils. All four events were recorded in the first 221 days of operation, and none have occurred since the rf excitation frequency for each SQUID was adjusted to avoid mutual resonances. Neverthe-

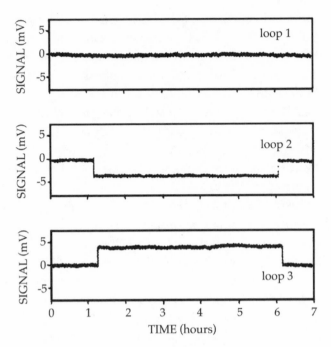

FIGURE 11.2. The calibration signal equivalent for one Dirac magnetic charge when the long toroidal coil is inserted into two loops of the apparatus. From Huber et al. (1990).

less, we have discarded the area contributed by adjacent-panel events, reducing our quoted sensing area to 1.1 m². (Huber et al. 1991)

The experimenters had eliminated a possible cause of spurious signals.

The researchers concluded that their data set an upper limit, which "is a factor of 2000 below the flux suggested by the single-candidate event seen with the prototype detector. *Based on this large factor and based on the noncoincident nature of the prototype detector, we conclude that the entire data set from the prototype detector which contains the single event should be discarded*" (Huber et al. 1990; emphasis added).

The demonstrated improvements in the experimental apparatus and analysis, including the coincidence requirements and the monitoring instruments, along with the failure to reproduce the original effect persuaded both the experimental group and the physics community that the original event was spurious and that magnetic monopoles had not been observed. Other monopole searches agreed.

In an interesting epilogue, Cabrera, in a talk at the University of Colorado (I was present), stated that the single-monopole candidate was made even less plausible by the fact that he and his group could artificially generate a similar signal with their toroidal coils. If they could do it, so could an intruder, and the laboratory was unoccupied when that event occurred. Although it was unlikely, it was possible that human intervention might have caused that signal. The third version of the experimental apparatus, which contained a motion sensor, eliminated that possibility in the last run, and would have done so, had it been present, in the prototype experiment. No good evidence for the existence of magnetic monopoles has been produced.

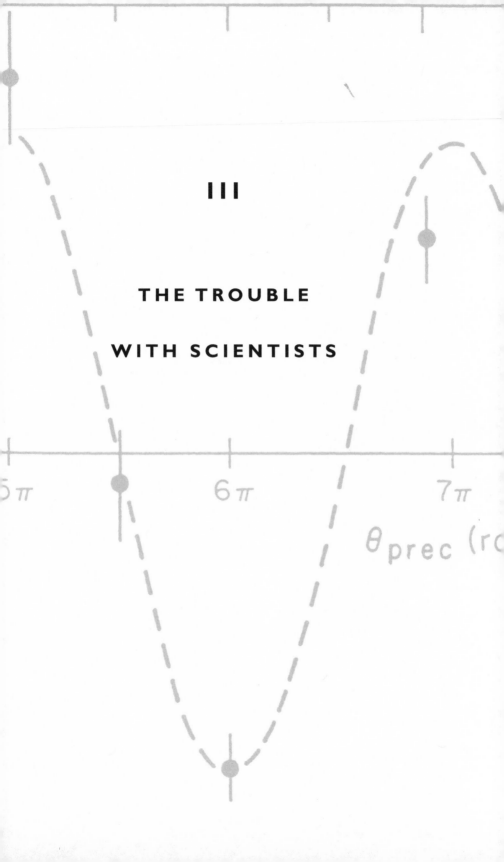

III

THE TROUBLE
WITH SCIENTISTS

(12)

Robert Millikan and the
Charge of the Electron

Robert Millikan's oil-drop experiments are justly regarded as a major contribution to 20th-century physics (Millikan 1911, 1913). They established the quantization of electric charge and the existence of a fundamental unit of charge and measured that unit of charge precisely. Earlier determinations of the charge of the electron had not established whether there was such a fundamental unit of electricity. This was because previous experiments, which used a cloud of charged water droplets and observed the motion of the cloud both under the influence of gravity and an electric field and under gravity alone, measured the total charge of the cloud and could not therefore demonstrate that the value obtained wasn't a statistical average. Millikan was able to perform all of his measurements on a single oil drop and avoid that difficulty. Examination of Millikan's laboratory notebooks reveals, however, that he was selective both in his choice of data and in his analysis procedure. The effects of this selectivity were small, however, and did not significantly affect Millikan's final value for e, the charge of the electron.

For more details, see Franklin (1981).

Nevertheless, the question remains of whether this well-known scientist modified both his data and his analysis procedure to achieve the desired result.

Millikan's Method

Millikan allowed a single charged oil drop to fall a known distance in air. He did not measure the time of fall from rest, but allowed the drop to fall freely for a short distance before it passed a crosshair, which signaled the start of the time measurement. Air resistance caused the drop to travel at a constant, terminal velocity. After the drop passed a second crosshair, which determined the time of fall at constant speed for the known distance between the crosshairs, an electric field was turned on. The charged oil drop then traveled upward at a different constant speed, and Millikan measured the time to ascend the same distance. These two time measurements allowed the determination of both the mass of the drop and its total charge.

The total charge on an oil drop is proportional to $[1/n(1/t_g + 1/t_f)]$, where n is the total number of charges on the drop in units of e, t_g is the time to fall a certain distance, and t_f is the time of ascent with the electric field on. The change in the charge of the oil drop can also be calculated. It is proportional to $[(1/\Delta n)(1/t'_f - 1/t_f)]$, where t'_f is the time of ascent after the charge on the drop has changed and Δn is the change in the charge of the drop in units of e.

Sample data sheets from Millikan's experiments are shown in figures 12.1 and 12.2. (These sheets are from Millikan's notebooks of 1911 and 1912, which are in the archives of the California Institute of Technology.* The results of that experiment were published in 1913.) The columns labeled G and F are the measurements of t_g and t_f, respectively. The average value of t_g and its reciprocal are given at the bottom of column G. To the right of column F are calculations of $1/t_f$ and of $[(1/\Delta n)(1/t'_f - 1/t_f)]$. Further to the right is the calculation of $[(1/n)(1/t_f + 1/t_g)]$. The top of the page gives the date; the number and time of the observa-

*All notebook references are from the R. A. Millikan Collection, California Institute of Technology. See Gunns and Goodstein (1975).

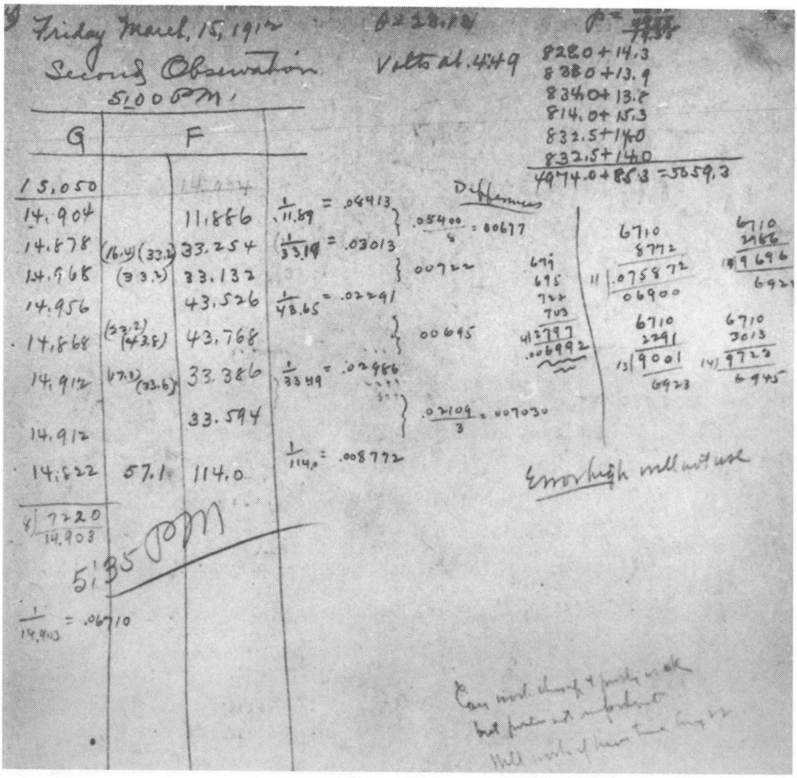

FIGURE 12.1. Millikan's data sheet for March 15, 1912 (second observation). The notation "Error high will not use," appears on the right. Courtesy California Institute of Technology Archives.

tion; the temperature θ; the pressure p; the voltage readings, which include the actual reading plus a correction; and the time at which the voltage was read. The data combined with the physical dimensions of the apparatus, the density of clock oil and of air, the viscosity of air, and the value of g are all that is required to calculate e.

Millikan's Results

Millikan could determine e from both the total charge of the drop and the changes in the charge. Not only did these values agree very well, but the average value obtained from different drops was also the same.

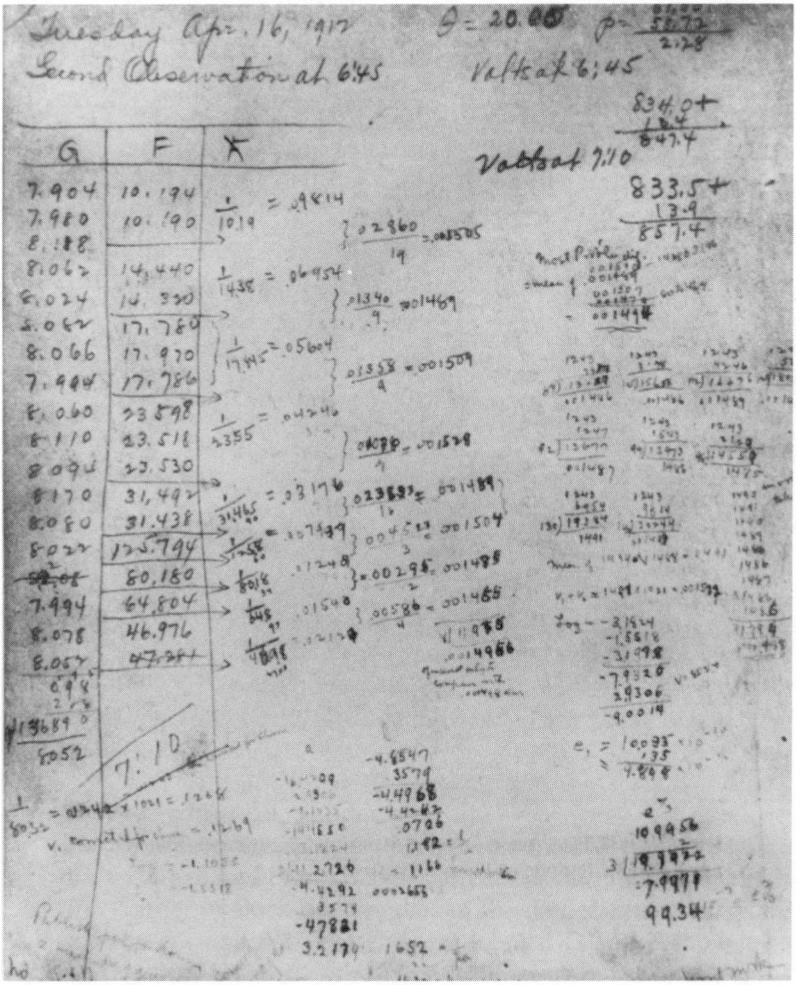

FIGURE 12.2. Millikan's data sheet for April 16, 1912 (second observation). The notation "Won't work" appears in the lower right-hand corner. Courtesy California Institute of Technology Archives

Millikan remarked (1911), "The total number of changes which we have observed would be between one and two thousand, and *in not one single instance has there been any change which did not represent the advent upon the drop of one definite invariable quantity of electricity or a very small multiple of that quantity.*" For Millikan, and for most of the physics community, these results established the quantization of charge. The value that Millikan found in 1911 for the fundamental unit of charge, the charge on the electron, was 4.891×10^{-10} esu (electrostatic units).

Following the completion of his 1911 paper, Millikan continued his oil-drop measurements. His intent was to improve both the accuracy and the precision of the measurement of e. He made improvements in his optical system and determined a better value for the viscosity of air. In addition, he took far more data in this second experiment. Millikan's new measurement gave a value of $e = (4.774 \pm 0.009) \times 10^{-10}$ esu. (Millikan's value for e differs from the modern value $e = (4.80320420 \pm 0.00000019) \times 10^{-10}$ esu. This is largely caused by the difference between the modern value for the viscosity of air and the one that Millikan used.) This value differs considerably from his 1911 value of 4.891×10^{-10} esu. Millikan stated (1913), "The difference between these numbers and those originally found by the oil-drop method, $e = 4.891$, was due to the fact that this much more elaborate and prolonged study had the effect of changing every one of the three factors η [the viscosity of air], A [related to the correction parameter in Stokes's Law], and d [the distance between the crosshairs], in such a way as to lower e. . . . The chief change, however, has been the elimination of faults of the original optical system."

In producing his new value of e, Millikan was selective both in the data he used and in his analysis procedure. In presenting his results in 1913, Millikan stated that the 58 drops under discussion had provided his entire set of data and added, "*It is to be remarked, too, that this is not a selected group of drops but represents all of the drops experimented upon during 60 consecutive days*," during which time the apparatus was taken down several times and set up anew." This is not correct. An examination of Millikan's notebooks for this period shows that he took data for this measurement from October 28, 1911, to April 16, 1912. My own count of the number of drops experimented on during this period is 175. Even if only those observations made after February 13, 1912 (the date of the first observation Millikan published), are counted, there are 49 excluded drops: that is, of 107 drops experimented on between February 13 and April 16, Millikan published the data for only 58.

It might appear that Millikan selectively analyzed his data to support his preconceptions about both charge quantization and the value of e.

Millikan's selectivity included the exclusion of all of the data of some single drops, the exclusion of some of the data within the data set for a single drop, and a choice of methods of calculation. It should be noted, however, that Millikan had far more data than he needed to improve the uncertainty in the measured value of e by approximately a factor of 10. Of the drops whose data he published, he used only those—23 out of a total of 58—that had a Stokes's law correction of less than 6% to calculate his final value of e. This was to guard against any effect of an error in that correction.

In experiments carried out before February 13, 1912, Millikan had labored to make his apparatus work properly. He was particularly worried about convection currents inside the device that could change the path of the oil drop. He made several tests on slow drops, for which convection effects would be most apparent. Millikan's comments on these tests are illuminating. On December 19, 1911, he remarked, "This work on a very slow drop was done to see whether there were appreciable convection currents. The results indicate that there were. Must look more carefully henceforth to tem[perature] of room." (The quotations are from Millikan's notebooks.) On December 20: "Conditions today were particularly good and results should be more than usually reliable. We kept tem very constant with fan, a precaution not heretofore taken in room 12 but found yesterday to be *quite* essential." On February 9, 1912, he disregarded his first drop because of uncertainty caused by convection; after the third drop he wrote, "This is good for so little a one but on these very small ones I must avoid convection still better." No further convection tests are recorded. It seems that by February 13 the device was working to Millikan's satisfaction, because he eventually published data from the very first drop recorded on that day. The data from 68 drops taken before February 13 were omitted from publication because Millikan was not convinced that his experimental apparatus had been working properly. It was not producing "good" data. Interestingly, the value for e that can be calculated from these 68 excluded events was $e = (4.75 \pm 0.01) \times 10^{-10}$ esu. This was, in fact, more precise than any other measurement available at the time. The data were, however, untrustworthy.

After this date, the apparatus was presumably working properly, unless Millikan's notebooks say otherwise. There are 107 drops in question, of which the data for 58 were published. Millikan made no calculation of e on 22 of the 49 drops whose data were not published. The most plausible explanation is that when he performed his final calculations in August 1912 they seemed superfluous: he did not need more drops for the determination of e when he already had so many.

The 27 events that Millikan did not publish and for which he calculated a value of e are more worrisome. Millikan knew the results he was excluding. Twelve of these were excluded from the published data because they seemed to require a second-order correction to Stokes's law. (The calculation of a value of e uses Stokes's law, which applies to the fall of an object in a continuous medium. To take into account the particulate character of air requires a correction to the law. Millikan used a first-order correction.) These 12 were very small drops, for which the value of Millikan's correction to Stokes's law was larger than one. This made Millikan's use of a perturbation series expansion of Stokes's law for those drops very questionable. There is no easy way to calculate the correction for such drops, so Millikan, having so much data, decided to exclude them. (I attempted, without success, to calculate a second-order correction to Stokes's law for these 12 drops. I found no consistent way to do so.) Of the remaining 15 calculated events, Millikan excluded 2 because the apparatus was not working properly, 5 because there was insufficient data to make a reliable determination of e, and 2 for no apparent reason. Six drops are left. One is anomalous; in the other five cases Millikan not only calculated a value for e, but compared it with an expected value. The four earliest events would have been placed in the group Millikan used to determine e. The only evident reason for rejecting these five events is that their values did not agree with his expectations. Including these events among the 23 that Millikan used to determine e would not significantly change the average value of e but would increase the statistical error of the measurement very slightly (see table 12.1).

The second drop of April 16, 1912, is quite anomalous among all of Millikan's data. (The data sheet for this drop is shown in figure 12.2.) It is also troublesome because it is among Millikan's most consistent measurements. Not only are the two methods of calculating e internally

TABLE 12.1

Comparison of Millikan's and Franklin's values of e

Published data for drops	values of $e \times 10^{10}$ esu			
	e(RM)	e(AF)	σ(RM)[a]	σ(AF)
First 23[b]	4.778[c]	4.773	±0.002	±0.004
All 58	4.780	4.777	±0.002	±0.003
Almost all drops[d]	4.781	4.780	±0.003	±0.003

Note: esu = electrostatic unit.

[a] Statistical error in the mean.

[b] These are the events that Millikan used to determine e.

[c] Although Millikan used a value $\mu = 0.001825$ for the viscosity of air in almost all of his calculations, in reporting his final value for e he used $\mu = 0.001824$. This accounts for the change from 4.774 to 4.778 in Millikan's final value. To make the most accurate comparison I used $\mu = 0.001825$ in all of my recalculations.

[d] This includes the data published for 58 drops, unpublished data for 25 drops measured after February 13, 1912, and some small corrections. For details, see Franklin (1981).

consistent, but they agree with each other very well. Millikan liked it: "Publish. Fine for showing two methods of getting v," he wrote in his notebook. My own calculation of e for this event gives a value $e = 2.810 \times 10^{-10}$ esu, or approximately $0.6e$. Millikan knew this. The comment "Won't work" appears in the lower right-hand corner of the data sheet. There were no obvious experimental difficulties that could explain the anomaly. Millikan remarked, "Something wrong with therm[ometer]," but there is no temperature effect that could explain a discrepancy of this magnitude. Millikan may have excluded this event to avoid giving Felix Ehrenhaft ammunition in the then-current controversy over the quantization of charge (see Holton 1978; Franklin 1981). In retrospect, Millikan was correct in excluding this drop. In later work William Fairbank Jr. and I found that Millikan's apparatus gave unreliable charge measurements when the charge on the drop exceeded a value of approximately $30e$. This drop had a charge of greater than $50e$, and the data were quite unreliable (Fairbank and Franklin 1982).

In addition to excluding the data from these five drops from publication, Millikan's cosmetic surgery touched 30 of the 58 published events. As shown in figures 12.1 and 12.2, Millikan made many measurements of the time of fall under gravity and of the time of ascent with the electric field on for each drop. In the data set for each of these 30 drops, Millikan excluded one or more (usually less than three) of these measurements. This group of 30 drops included several of the 23 drops Millikan used in his final determination of e as well some of the 35 drops for which data were published but not used. My recalculation of these events, using all of the data for each drop, gives results little different from Millikan's. Millikan's exclusion of these measurements was not based on the value of e he obtained for the drops, because, in general, Millikan did not include these measurements in his calculations and therefore did not know their effect.

There are two ways to calculate e from the oil-drop data. The first uses the total charge on the drop, and the second uses the changes in that charge. Millikan claimed that he had used the first method exclusively because the large number of measurements of t_g provided a more accurate determination of e. In at least 19 of the 58 published events, however, he used either the average value of the two methods, some combination of the two that is not a strict average, or the second method alone. (Figure 12.1 shows the data sheet for an event that Millikan excluded because he thought the difference in the value of e obtained by the two different methods was too large. He wrote on the sheet, "Error high will not use.") In general, the effects are small, and the result of his tinkering, once again, is to reduce the statistical error very slightly rather than to change the mean value of e. The effect of all of Millikan's selectivity is shown in table 12.1. This includes Millikan's results along with my own recalculation of Millikan's data. The results of his selectivity are quite small.

What can be concluded from this episode? Millikan intended to establish the quantization of charge and to measure the fundamental unit more accurately and precisely than had been done previously. It is apparent that he succeeded: there is no reason to disagree with his assessment that, in 1913, there was "no determination of e . . . by any other method which does not involve an uncertainty at least 16 times as great as that represented in these measurements." His apparently arbitrary exclusion of five drops for which he had calculated e had an utterly negligible effect on his final result. Because Millikan knew the value of e obtained from the events he was excluding, he also knew that the effect of the exclusions, and of his selective calculations, on his final result was small. The effect of Millikan's selectivity was to reduce the statistical uncertainty of his final result very slightly. It had no significant effect on the final value of e (see table 12.1). Almost all of the uncertainty in Millikan's final value was due to systematic effects— uncertainty in the distance between the plates, uncertainty in the voltage, and so on. It is unclear why he was so worried about the statistical uncertainty.

Nevertheless there is strong evidence that Millikan tailored his cuts and his analysis procedure to obtain the result he wanted. Several of these cuts seem quite legitimate: the exclusion of the early drops because he was not sure that the experimental apparatus was working properly, the exclusion of some data within the set for a drop, and the exclusion of later events because he simply did not need them for his calculations. The exclusion of drops for which he calculated a value of e and could thus select the value he wanted, as well as his choice of calculational method, is not justified. The fact of Millikan's modifications was unavailable to the physics community. Millikan's questionable selectivity remained private. In fact, his statements in his 1913 paper that the drops were not selected and that he used only one method of calculation seem to have been designed to conceal that selectivity. Science has safeguards against procedures such as Millikan's, which, in less sure hands, could easily have unfortunate results. Replication is the safety mechanism for a case like this. The value of e was an important physical quantity. It was used in the calculation or determination of many important physical constants—Avogadro's number, the Rydberg constant, and others. There were many repetitions of Millikan's measurement. It is rare that an important physical quantity is measured only once. Had Millikan's selectivity grossly affected his measured value of e, a discrepancy with later measurements would certainly have appeared.

(13)

The Early Searches for Gravity Waves

Divergent views exist on the nature of science and experiment: the social contructivist view holds that science is influenced, if not determined, by social, political, and economic context, and the rationalist view (my own) regards science as a reasonable enterprise based on valid experimental evidence and on reasoned and critical discussion.

Collins and the Experimenters' Regress

Harry Collins is well known for his skepticism concerning both experimental results and evidence. He calls his argument the experimenters' regress (Collins 1985, chap. 4): What scientists take to be a correct result is one obtained with a good—that is, properly functioning—experimental apparatus. But a good experimental apparatus is simply

For a more detailed discussion, see Franklin (1994). For Collins's view, see Collins (1985; 1994). For an account by one of the original experimenters, see Levine (2004).

one that gives correct results. Collins avers that no formal criteria can be applied to decide whether or not an experimental apparatus is working properly. In particular, he argues that calibrating an experimental apparatus by using a surrogate signal cannot provide an independent reason for considering the apparatus to be reliable.

In Collins's view, the regress is eventually broken by negotiation within the appropriate scientific community, a process driven by factors such as the career, social, and cognitive interests of the scientists, but one that is not decided by what might be called epistemological criteria, or reasoned judgment. Thus Collins concludes that his regress raises serious questions concerning both experimental evidence and its use in evaluating scientific hypotheses and theories. Indeed, if no way out of the regress can be found, then he has a point.

Collins presents his strongest candidate for an example of the experimenters' regress in actual scientific practice in his history of the early attempts to detect gravitational radiation, or gravity waves. In this case, the scientific community was forced to compare Joseph Weber's claims that he had observed gravitational waves with the reports of six other experiments that failed to detect them. On the one hand, Collins argues that the decision between these conflicting experimental results could not be made on epistemological or methodological grounds. He claims that the six negative experiments could not legitimately be regarded as replications and hence become less compelling. Collins offers two arguments concerning the difficulty, if not the virtual impossibility, of replication. The first is philosophical and concerns what it means to replicate an experiment and in what way the replication is similar to the original experiment. A rough-and-ready answer from supporters of the scientific process is that the replication measures the same physical quantity. Whether or not it in fact does so can be asserted on reasonable grounds.

Collins's second argument is pragmatic. In practice it is often difficult to get an experimental apparatus, even one known to be similar to another, to work properly. Collins illustrates this with his account of Harrison's attempts to construct two versions of a TEA (transverse excited atmospheric) laser (1985, 51–78). Despite the fact that Harrison had previous experience with such lasers, and had excellent contacts with experts in the field, he had great difficulty in building the lasers. Hence the difficulty of replication.

Ultimately Harrison found errors in his apparatus, and once these were corrected, the lasers operated properly. As Collins acknowledges (p. 84), "In the case of the TEA laser the circle was readily broken. The ability of the laser to vaporize concrete, or whatever, comprised a universally agreed criterion of experimental quality. There was never any doubt that the laser ought to be able to work and never any doubt about when one was working and when it was not."

Although Collins seems to consider that Harrison's problems with replication illuminate the episode of gravity waves, support the experimenters' regress, and cast doubt on experimental evidence in general, his argument fails to convince. As Collins admits, the replication was clearly demonstrable. It is unclear what role Collins thinks this episode plays in his argument.

On the other hand, Weber's apparatus, precisely because the experiments used a new type of apparatus to try to detect a hitherto-unobserved phenomenon, could not be subjected to standard calibration techniques. I argued previously (1994, 1995b) that the gravity wave experiment is not at all typical of physics experiments. In most experiments, the adequacy of the surrogate signal used in the calibration of the experimental apparatus is clear and unproblematic. In cases where it is questionable, considerable effort is devoted to establishing the adequacy of that surrogate signal. Although Collins chose an atypical example, the questions he raises about calibration in general and about this particular episode of gravity wave experiments should be answered. Even in this extremely complex and atypical case of experimental uncertainty, scientists had good orthodox epistemic reasons for arriving at their conclusions about whether gravity waves had really been detected.

Collins's Account of Gravity Wave Detectors

Collins begins his illustration of the experimenters' regress with a discussion of the original apparatus developed by Joseph Weber (figure 13.1), which later became standard. Weber used a massive aluminum

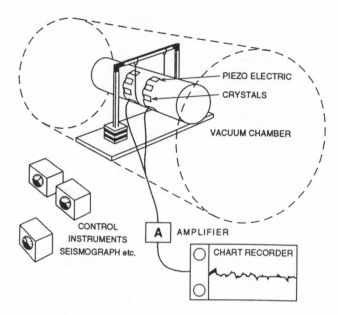

FIGURE 13.1. A Weber-type gravity wave detector. From Collins (1985).

alloy bar, or antenna, which was supposed to oscillate when struck by gravitational radiation. This device is often referred to as a Weber bar. The gravity waves it is intended to detect are predicted by Albert Einstein's general theory of relativity. Just as an accelerated electrically charged particle will produce electromagnetic radiation (light, radio waves, etc.), so should an accelerated mass produce gravitational radiation—gravity waves. Such radiation can be detected by the oscillations produced in a large mass when it is struck by gravity waves. Because the gravitational force is far weaker than the electromagnetic force, a large mass must be accelerated to produce a detectable gravity wave signal. The ratio of the gravitational force between the electron and the proton in the hydrogen atom and the electrical force between them is 4.38×10^{-40}, a small number indeed. The difficulty of detecting a weak signal is at the heart of this episode.

The oscillation was to be detected by observing the amplified signal from piezo-electric crystals attached to the antenna. The signals were expected to be quite small—the gravitational force is much weaker than the electromagnetic force—and the bar had to be insulated from other sources of noise, such as electrical, magnetic, thermal, acoustic, and seismic forces. Because the bar was at a temperature different from absolute zero, thermal noise could not be avoided, so Weber set a threshold for pulse acceptance that was larger than the signals expected from thermal noise. In 1969, Weber claimed to have detected approximately seven above-threshold pulses per day caused by gravitational radiation, in excess of those expected from thermal noise.

Because Weber's reported rate was far greater than that expected from calculations of cosmic events (by a factor of more than 1,000), his early claims were met with skepticism. During the late 1960s and early 1970s, however, Weber introduced several modifications and improvements that increased the credibility of his results (Weber et al. 1973). He claimed that above-threshold peaks had been observed simultaneously in two detectors separated by 1,000 miles. It was extremely unlikely that such coincidences were due to random thermal fluctuations. In addition, he reported a 24-hr periodicity in his peaks, the sidereal correlation, that indicated a single source for the radiation, perhaps near the center of our galaxy. These results increased the plausibility of his claims sufficiently that by 1972 three other experimental groups had not only built detectors but also reported results. However, none were in agreement with Weber.

Collins notes that after 1972, Weber's claims were less and less favored. During 1973 three different experimental groups reported negative results, and subsequently these groups, as well as three others, reported further negative results. No corroboration of Weber's results was reported during this period. Although in 1972 approximately a dozen groups were involved in experiments aimed at checking Weber's findings, by 1975 no one, except Weber himself, was still working on that particular problem. Weber's results were regarded as incorrect.

Different scientists offered different reasons for their rejection of Weber's claims, and not all of the researchers engaged in the pursuit agreed about their importance. Between 1972 and 1975 it was discovered that Weber had made several serious errors in his analysis of his data. His computer program for analyzing the data contained an error, and his statistical analysis of residual peaks and background was questioned and thought to be inadequate. Weber also claimed to find coincidences between his detector and another distant detector when, in fact, the tapes used to provide the coincidences were actually recorded more than 4 hr apart. Weber had found a positive result where even he did not expect one. Others cited the failure of Weber's signal-to-noise ratio to improve, despite his "improvements" to his apparatus. In addition, the sidereal correlation disappeared. This was perhaps just as well, because it had been pointed out that a sidereal effect should have a 12-hr period. The earth should not absorb gravity waves. Perhaps most striking were the uniformly negative results obtained by six other groups. Collins argues, however, that these reasons were insufficient to reject Weber's results.

If Collins is correct that the negative evidence provided by the replications of Weber's experiment—the application of epistemological criteria, combined with Weber's acknowledged errors—is insufficient to explain the rejection of Weber's results, then he must provide another explanation. Collins invokes the experimenters' regress: if the regress is a real problem in science, then scientists should disagree about what constitutes a good detector, and this is exactly what his fieldwork shows. Collins presents several excerpts from interviews with scientists working in the field that show differing opinions on the quality of detectors. At the beginning, when scientists were responding positively to Weber's experiments, they offered different reasons for believing in Weber's claims. Some were impressed by the coincidences between two separated detectors; others cited the fact that the coincidence disappeared when one detector signal was delayed relative to the other or approved of Weber's use of the computer for analysis. But not everyone agreed, and 3 years later when the majority view had shifted against Weber, scientists again gave different reasons for rejecting his program. To Collins, these differing opinions demonstrate the lack of any consensus over formal criteria for the adequacy of gravitational wave detectors.

Collins notes that after 1975 scientists stopped pursuing other possible explanations of the conflict between Weber's results and those of his critics. These included not only prosaic differences in the detectors —that is, piezo-electric crystals compared with other strain detectors, the antenna material, and the electronics—but also more theoretical hypotheses such as the invocation of a new fifth force. There was even speculation in some quarters that the gravity wave findings were the result of random mistakes, deliberate lies, or self-deception or even that they could be explained by psychic forces. Yet by 1975 all of these alternative explanations of the discordant results of Weber and his critics had been dropped. It was simply accepted that Weber had made systematic errors. Collins suggests that this was not a necessary conclusion and that scientists might have reasonably investigated the other, more radical possibilities. Collins proposes that the key factor in the formation of a premature consensus was the social impact of the negative evidence provided by scientist Q. (Any reader of the literature will easily identify Q as Richard Garwin.) Collins argues that it was not so much the intrinsic power of Q's experimental result but rather his forceful and persuasive presentation of that result and his careful analysis of thermal noise in an antenna that turned the tide.

In the course of presenting his case that there were insufficient epistemic grounds for rejecting Weber's reports, Collins deals with the attempt to break the experimenters' regress by the use of experimental calibration. In this episode, most experimenters calibrated their gravity wave detectors by injecting a pulse of known acoustical energy at one end of their antenna and measuring the output of their detector. This served to demonstrate that the apparatus could detect energy pulses and provided a measure of the sensitivity of the apparatus.

According to Collins, Weber was initially reluctant to calibrate his own antenna acoustically, but eventually did so. His results included, however, a quite different method of analyzing the output pulses. He used a nonlinear energy algorithm, which was sensitive only to the amplitude of the signal, whereas his critics used a linear algorithm that was sensitive to both the phase and the amplitude of the signal. The critics argued that it was possible to show rigorously and mathematically that the linear algorithm was superior in detecting pulses. The issues of the calibration of the apparatus and the method of analysis used were inextricably tied together. When the calibration was done on Weber's apparatus, it was found that the linear algorithm was 20 times better than Weber's nonlinear algorithm at detecting the calibration signal. For the critics, this established the superiority of their detectors. Weber did not agree. He argued that the analysis and calibration applied only to short pulses, those expected theoretically and used in the calibration, whereas the signal he was detecting had a length and shape that made his method superior.

Collins concludes,

> The anomalous outcome of Weber's experiments could have led toward a variety of heterodox interpretations with widespread consequences for physics. They could have led to a schism in the scientific community or even a discontinuity in the progress of science. Making Weber calibrate his apparatus with the electrostatic pulses was one way in which his critics ensured that gravitational radiation remained a force that could be understood within the ambit of physics as we know it. They ensured physics' continuity—the maintenance of links between past and future. Calibration is not simply a technical procedure for closing debate by providing an external criterion of competence. In so far as it does work this way, it does so by controlling interpretive freedom. It is the control on interpretation which breaks the circle of the experimenters' regress, not the "test of a test" itself. (1985, 105–6)

Collins states that the purpose of his argument is to demonstrate that science is uncertain. He grants, however, "For all its fallibility, science is the best institution for generating knowledge about the natural world that we have."

Although I agree with Collins concerning the fallibility of science and its great value, I find serious problems with his argument. These are particularly important because the regress argument, despite Collins's disclaimer, casts doubt on experimental evidence and on its use in science and therefore on the status of science as knowledge.

Collins's argument can be briefly summarized as follows: There are no rigorous independent criteria for either a valid result or a good experimental apparatus; all attempts to evaluate the apparatus are dependent on the outcome of the experiment. This leads to the experimenters' regress, in which a good detector can be defined only by its obtaining the correct outcome, whereas a correct outcome is one obtained with a good detector. In practice, the regress is broken by negotiation within the scientific community, but the decision is not based on anything that might be called epistemological criteria. This casts doubt on not only the certainty of experimental evidence, but its very validity. Thus, experimental evidence cannot provide grounds for scientific knowledge.

An Alternative History of Gravity Wave Detection

Collins might correctly argue that the case of gravity wave detectors is a special one, in which a new type of apparatus was used to try to detect a hitherto-unobserved quantity. I agree. But I do not agree that arguments could not be presented concerning the validity of the results or that the relative merits of two results could not be evaluated, independent of the outcome of the two experiments. The regress can be broken by reasoned argument. The published record gives the details of that reasoned argument.*

In the controversy surrounding the early attempts to observe gravity waves, the crucial question was not what constituted a good gravity wave detector; all the experiments used variants of the Weber antenna. Rather, the questions were whether the detector was operating properly and whether the data were being analyzed correctly. The discordant results reported by Weber and his critics and the disagreements about how best to resolve the differences are not unusual occurrences in the history of physics, particularly at the beginning of an experimental investigation of a phenomenon. Collins correctly concludes on the basis of his interviews that Weber's critics did not always agree about the force of particular arguments, but this does not mean that by the end of the controversy each did not have compelling reasons for believing that Weber was wrong. Understanding how the scientists arrived at their conclusions requires an examination of the full history of the episode, as given in published papers, conference proceedings, and public let-

*In this section, I rely primarily on a panel discussion on gravitational waves that took place at the Seventh International Conference on General Relativity and Gravitation (GR7), Tel Aviv University, June 23–28, 1974. The panel included Weber and three of his critics, Tyson, Kafka, and Drever, and included not only papers presented by the four scientists but also discussion, criticism, and questions. Almost all of the important and relevant arguments concerning the discordant results were discussed. The proceedings were published as Shaviv and Rosen (1975).

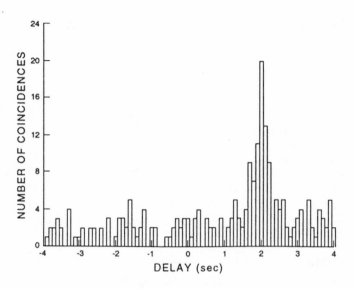

FIGURE 13.2. A plot showing the calibration pulses for the Rochester–Bell Laboratory collaboration. The peak caused by the calibration pulses is clearly seen. From Shaviv and Rosen (1975).

ters. These paint a picture of overwhelming evidence against Weber's result and show that the final decision, although not governed by an algorithm, was unquestionably reasonable and based on epistemological criteria.

It is difficult to determine whether there is a signal in a gravitational wave detector or whether two such detectors have fired simultaneously. There are several problems. There are energy fluctuations in the bar due to thermal, acoustic, electrical, magnetic, and seismic noise. When a gravity wave strikes the antenna, its energy is added to the existing energy. This may change the amplitude or the phase, or both, of the signal emerging from the bar. Simply observing a larger signal from the antenna after a gravitational wave strikes it will not suffice. This difficulty informs the discussion of which was the best analysis procedure to use.

The nonlinear, or energy, algorithm preferred by Weber was sensitive only to changes in the amplitude of the signal. The linear algorithm, preferred by everyone else, was sensitive to changes in both the amplitude and the phase of the signal. Weber preferred the nonlinear

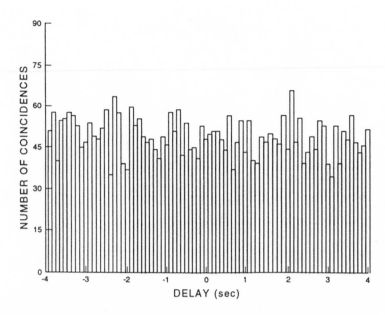

FIGURE 13.3. A time-delay plot for the Rochester–Bell Laboratory collaboration, made with the nonlinear algorithm. No sign of any calibration peak is seen. From Shaviv and Rosen (1975).

procedure because it resulted in the output of several pulses exceeding the threshold for each input pulse to his detector. Weber admitted, however, that the linear algorithm, preferred by his critics, was more efficient at detecting calibration pulses. Results on the superiority of the linear algorithm for detecting calibration pulses were reported by both Kafka and Anthony Tyson. Tyson's results for calibration pulse detection are shown for the linear algorithm in figure 13.2 and for the nonlinear algorithm in figure 13.3. The peak for the linear algorithm is clear, whereas the peak for the nonlinear procedure is not apparent. (The calibration pulses were inserted periodically during data-taking runs. The peak was displaced by 2 sec by the insertion of a time delay, so that the calibration pulses would not mask any possible real signal, which was expected at zero time delay.)

Nevertheless, Weber preferred the nonlinear algorithm. His reason was that this procedure gave a more significant signal than did the linear one. This is illustrated in figure 13.4, in which Weber's data analyzed with the nonlinear algorithm are presented in the top panel, and

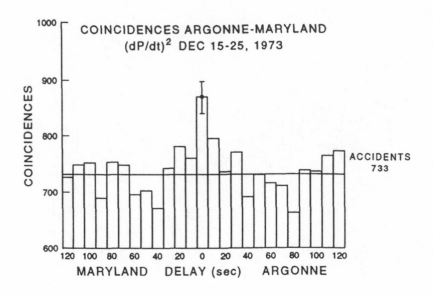

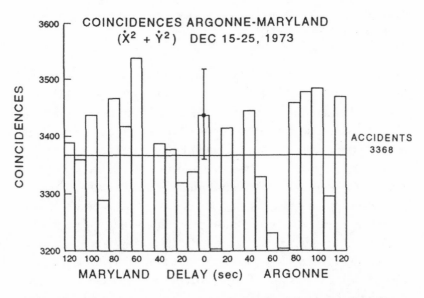

FIGURE 13.4. Weber's time-delay data for the Maryland-Argonne collaboration for the period December 15–25, 1973. Top, nonlinear algorithm; bottom, linear algorithm. The zero-delay peak is seen only with the nonlinear algorithm. From Shaviv and Rosen (1975).

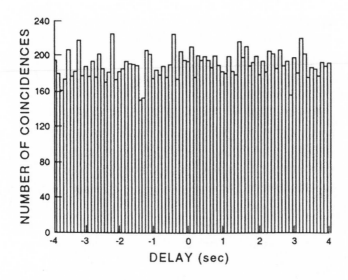

FIGURE 13.5. A time-delay plot for the Rochester–Bell Laboratory collaboration, made with the linear algorithm. No sign of a zero-delay peak is seen. From
Shaviv and Rosen (1975).

the same data analyzed with the linear procedure are presented in the
bottom panel. Weber was choosing the method that gave a positive
result. If anyone was "regressing," it was Weber. However, contrary to
Collins's account, the rest of the scientific community was sharply critical of this maneuver.

Weber's critics objected to his failure to calibrate his apparatus.
Nonetheless, they accommodated Weber's position as much as possible
by analyzing their own data with both algorithms. The linear algorithm
was clearly superior for analyzing the calibration pulses, but if it either
masked or failed to detect a real signal, then using the nonlinear algorithm on their data should have produced a clear signal. But none appeared. Typical results are shown in figures 13.3 and 13.5. Figure 13.3 gives
Tyson's data analyzed with the nonlinear algorithm and shows not only
no calibration peak but also no signal peak at zero time delay. It is
quite similar to the data analyzed with the linear algorithm shown in
figure 13.5. (Note that for this data run, no calibration pulses were inserted.) Collins does not discuss the fact that Weber's critics exchanged

both data and analysis programs and that they analyzed their own data with Weber's preferred nonlinear analysis algorithm and failed to find a signal. This is documented in the published record and demonstrates that epistemological criteria figured in the evaluation of the discordant experimental results.

Weber had an answer ready. He suggested that although the linear algorithm was better for detecting calibration pulses, which were short, the real signal of gravitational waves was a longer pulse than most investigators thought. He argued that the nonlinear algorithm that he used was better at detecting these longer pulses. The critics did think that gravitational radiation would be produced in short bursts. Still, if the signal was longer, a positive result would be expected to appear when the critics' data were processed with the nonlinear algorithm. It did not. Ronald Drever also reported that he had looked at the sensitivity of his apparatus with arbitrary waveforms and pulse lengths. Although he found a reduced sensitivity for longer pulses, he analyzed his data to look explicitly for such pulses. He found no effect. He also found no evidence for gravity waves when he used the short-pulse (linear) analysis.

Collins claims that the experiments performed by the critics could not be legitimately considered to be replications; however, there was considerable cooperation among the various groups. They exchanged both data tapes and analysis programs, and this led to the first of several questions about possible serious errors in Weber's analysis of his data. David Douglass first pointed out that there was an error in one of Weber's computer programs.

> The nature of the error was such that any above-threshold event in antenna A that occurred in the last or the first 0.1 sec time bin of a 1000 bin record is erroneously taken by the computer program as in coincidence with the next above-threshold event in channel B, and is ascribed to the time of the later event. Douglass showed that in a four-day tape available to him and included in the data of [Weber's 1973 paper], nearly all of the so-called "real" coincidences of 1–5 June (within the 22 April to 5 June 1973 data) were created individually by this simple programming error. Thus not only some phenomenon besides gravity waves *could*, but in fact *did* cause the zero-delay excess coincidence rate. (Garwin 1974)

Weber admitted the error but did not agree with the conclusion. He claimed that even after the error was corrected, there was a positive signal. (This claim was never corroborated in the published literature.) It is clear, however, that this error raised doubts about the correctness of Weber's results.

Another serious question was raised concerning Weber's use of varying selection thresholds in the analysis of his data. Weber's critics used a single threshold to analyze all of their data. They accused Weber of varying his threshold to maximize his final result. Garwin used a computer simulation to show that such a procedure could produce a positive result that was merely an artifact of the selection procedure. Kafka also showed, with his own data, that varying the selection procedure could, on occasion, create a positive signal. James Levine and Garwin also questioned whether Weber's apparatus could have produced the signal he reported. They used a computer simulation to argue that the signal should have been far broader than the one Weber published. Weber denied both charges. He did not specify his method of data selection for his histogram, however. In particular, he did not state that all of the results presented in a particular histogram had been obtained with the same threshold.

Weber reported another odd result.

First, Weber has revealed at international meetings that he had detected a 2.6-standard deviation excess in coincidence rate between a Maryland antenna [Weber's apparatus] and the antenna of David Douglass at the University of Rochester. Coincidence excess was located not at zero time delay but at "1.2 seconds," corresponding to a 1-sec intentional offset in the Rochester clock and a 150-millisecond clock error. At CCR-5, Douglass revealed, and Weber agreed, that the Maryland Group had mistakenly assumed that the two antennas used the same time reference, whereas one was on Eastern Daylight Time and the other on Greenwich Mean Time. Therefore, the "significant" 2.6-standard deviation excess referred to gravity waves that took four hours, zero minutes and 1.2 seconds to travel between Maryland and Rochester. (Garwin 1974)

Weber answered that he had never claimed that the 2.6-standard-deviation effect he had reported was a positive result. But by producing a statistically significant result where none was expected, Weber had certainly cast doubt on his analysis procedures.

The critics' results were clearly far more credible than Weber's. They had checked their results by independent confirmation, which included sharing of data and analysis programs. They had eliminated a plausible source of error, that of the pulses being longer than expected, by analyzing their results with the nonlinear algorithm and by looking for such long pulses. They had also calibrated their apparatuses by injecting pulses of known energy and observing the output. Contrary to Collins's assertions, physicists made a reasoned judgment when they rejected Weber's results and accepted those of his critics. Although no formal rules were applied—if four errors, rather than three, are found, the results lack credibility, or if there are five, but not six, conflicting results, the work is still credible—the procedure clearly followed the norms of scientific experimentation.

Collins's account of Garwin's role (scientist Q) is also questionable. Although Garwin was a major figure in the controversy and presented strong and forceful arguments against Weber's result, the same arguments were being made at the time by other scientists. At the 1974 Conference on General Relativity, for example, in the panel discussion on gravity waves, Garwin's experiment was mentioned only briefly, and although the arguments about Weber's errors and analysis were made, they were not attributed to the absent Garwin.

A point that should be emphasized is that although calibration—and its success or failure—played a significant role in the dispute, it was not decisive. Other arguments were both needed and provided. In most cases, failure to detect a calibration signal would be a decisive reason for rejecting an experimental result. In this case, however, it was not, because the scientists involved seriously considered the question of whether an injected acoustic energy pulse was an adequate surrogate for a gravity wave. Their doubts about its adequacy led to the variation in analysis procedures and to the search for long pulses.

Collins is mistaken in his claim that the experimenters' regress played a role in this episode. He conflates the difficulty of getting an experiment to work with the problem of demonstrating that it is working properly. Epistemological criteria were reasonably applied to decide between Weber's results and those of his critics. Although calibration was not decisive in the case of gravity wave detectors, nor should it have been, it is often a legitimate and important factor and may even be decisive in determining the validity of an experimental result. The arguments that replication was impossible and that criteria to decide the validity of experimental results were absent are unsupported. The history of gravity wave detectors provides no proof for the experimenters' regress.

At present gravity waves have not been detected either with Weber-bar antennae or the newer interferometer, in which the gravitational radiation will have a differential effect on the two arms of the device and thus change the observed interference pattern. The radiation has not been detected even though current detectors are several orders of magnitude more sensitive than those in use in 1975.

Gravity waves have, however, been observed indirectly. They have been detected by measuring the change in orbital period of a binary pulsar. Such a binary system should emit gravitational radiation, thereby losing energy and decreasing the orbital period. This effect was initially measured by using the results of R. Hulse and Joseph Taylor (1975), which provided the initial measurement of the period, and of J. Weisberg and Taylor (1984), which measured the period later. The measured change in the period was $(-2.40 \pm 0.09) \times 10^{-12}$ s s^{-1}, in excellent agreement with the theoretical prediction of $(-2.403 \pm 0.002) \times 10^{-12}$ s s^{-1}. "As we have pointed out before most relativistic theories of gravity other than general relativity conflict strongly with our data, and would appear to be in serious trouble in this regard. It now seems inescapable that gravitational radiation exists as predicted by the general relativistic quadrupole formula" (Weisberg and Taylor 1984). More recent measurements and theoretical calculations give $(2.427 \pm 0.026) \times 10^{-2}$ s s^{-1} (measured) and $(2.402576 \pm 0.000069) \times 10^{-12}$ s s^{-1} (theoretical). If the general relativity theory is correct, Weber should not have observed a positive result. Weber's original signal was 1,000 times larger than that predicted by general relativity.

(14)
Atomic-Parity Violation

DO MUTANTS DIE OF NATURAL CAUSES?

Andrew Pickering's *Constructing Quarks* (1984b) is an early product of the social constructivist attempt to portray scientific theories as products of social negotiation, driven by factors such as the career, social, and cognitive interests of their formulators. In this view, the decision about whether to accept a theory is not decided by evidence and what might be called epistemological criteria, or reasoned judgment. Pickering says in the introduction to this richly detailed history of high-energy physics that he will try to offer a mirror image of the standard realist accounts in which "experiment is seen as the supreme arbiter of theory" (p. 5). "The scientist legitimates scientific judgments by reference to the state of nature; I attempt to understand them by reference to the cultural context in which they are made" (p. 8). During much of the two decades covered in Pickering's book, elementary-particle theoreticians were casting about in many directions, and experimenters were correspondingly unsure which experiments to undertake and which parameters to measure. Pickering's account of these uncertain periods, based

For more details, see Franklin (1990, chap. 8).

in part on extensive interviews with key players, is both fascinating and unforced. It is only when scientists feel that they are getting some answers and their views start to converge that Pickering's history starts to diverge from what his informants are reporting.

Pickering focuses on the period following the "November revolution" of 1974, when the Weinberg-Salam (W-S) theory quickly became the standard model of electroweak interactions. (Steven Weinberg, Abdus Salam, and Sheldon Glashow received the Nobel Prize for Physics in 1979.) Unimpressed by the widespread experimental confirmations of this theory, Pickering remarks, "Quite simply, particle physicists accepted the existence of the neutral current [one of the predictions of the W-S theory] because they could see how to ply their trade more profitably in a world in which the neutral current was real" (1984a). However, Pickering takes very seriously the early experiments on atomic-parity violation, which were anomalous for the W-S unified theory of electroweak interactions. These experiments, performed at Oxford University and the University of Washington and published in 1976 and 1977, measured the parity-nonconserving effect in atomic bismuth. The results disagreed with the predictions of the W-S theory; another parity-violating experiment, performed at the Stanford Linear Accelerator Center (SLAC) in 1978, which investigated the scattering of polarized electrons from deuterons, confirmed the theory.

In a section entitled "The Slaying of Mutants," Pickering argues that the scientific community simply regarded the Oxford and Washington experiments as mutants and didn't take them seriously. He claims that by 1979 the high-energy physics community accepted the W-S theory as established, even though "there had been no *intrinsic* change in the status of the Washington-Oxford experiments" (1984b, 301). In Pickering's view, "Particle physicists *chose* to accept the results of the SLAC experiment, *chose* to interpret them in terms of the standard model (rather than some alternative which might reconcile them with the atomic physics results) and therefore *chose* to regard the Washington-Oxford experiments as somehow defective in performance or interpretation" (1984b, 301).

The implication seems to be that these choices were made so that the experimental evidence would be consistent with the standard model and that there weren't good, independent reasons for them. In short,

Pickering argues that the mutants died not from natural causes but because they didn't fit in with the career interests of the opportunistic scientific community. To be fair, Pickering is discussing the adjustment of theoretical and experimental research practice. However, his meaning is essentially agreement between experiment and theory. I believe his judgment on the relative evidential weight of the two different experiments is incorrect.

Pickering regards this episode as support for his view that "there is no obligation upon anyone framing a view of the world to take account of what twentieth century science has to say" (1984b, 413). He thus denies that science is a reasonable enterprise based on valid experimental or observational evidence. (Pickering denies this allegation. For different points of view, see Franklin 1990, chap. 8, 1991, 1993a; Ackermann 1991; Lynch 1991; Pickering 1991).

The Washington and Oxford Experiments

The search for atomic-parity violation arose from the attempt to test the W-S unified theory of electromagnetic and weak interactions in a new domain. The model predicted that weak neutral-current effects would be seen in the interactions of electrons with hadrons, the strongly interacting particles. The effect would be quite small compared with that of the dominant electromagnetic interaction but could be distinguished from it by the fact that it violated parity conservation. Thus a demonstration of such a parity-violating effect and a measurement of its magnitude would test the W-S theory.

One such predicted effect was the rotation of the plane of polarization of polarized light when it passed through bismuth vapor. Such a rotation is possible only if parity is violated. This was the experiment performed by the Oxford and Washington groups. They both used bismuth vapor but used light corresponding to different transitions in bismuth, $\lambda = 648$ nm (Oxford) and $\lambda = 876$ nm (Washington). They published a joint preliminary report noting, "We feel that there is sufficient interest to justify an interim report" (Baird et al. 1976). They reported values for R, the parity-violating parameter, of $R = (-8 \pm 3) \times 10^{-8}$ (Washington) and $R = (+10 \pm 8) \times 10^{-8}$ (Oxford). "We conclude from the two experiments that the optical rotation, if it exists, is smaller than the values -30×10^{-8} and -40×10^{-8} predicted by the Weinberg-Salam model plus the atomic central field approximation."

Pickering offers this interpretation:

> The caveat to this conclusion was important. Bismuth had been chosen for the experiment because relatively large effects were expected for heavy atoms, but when the effect failed to materialise a drawback of the choice became apparent. To go from the calculation of the primitive neutral-current interaction of electrons with nucleons to predictions of optical rotation in a real atomic system it was necessary to know the electron wavefunctions, and in a multi-electron atom like bismuth these could only be calculated approximately. Furthermore, these were novel experiments and it was hard to say in advance how adequate such approximations would be for the desired purpose. Thus in interpreting their results as a contradiction of the Weinberg-Salam model the experimenters were going out on a limb of atomic theory. Against this they noted that four independent calculations of the electron wavefunctions had been made, and that the results of these calculations agreed with one another to within twenty-five percent. This degree of agreement the experimenters found "very encouraging" although they conceded that "Lack of experience of this type of calculation means that more theoretical work is required before we can say whether or not the neglected many-body effects in the atomic calculation would make R [the parity-violating parameter] consistent with the present experimental limits." (1984b, 295–96)

Pickering attributes all the uncertainty in the comparison between experiment and theory to the theoretical calculations and none to the experimental results themselves.

The comparison was even more uncertain than Pickering implies and included uncertainty in the experimental results. The experimenters reported that the "quoted statistical error represents 2 s.d. [standard deviations]. There are, however, also systematic effects which we believe do not exceed $\pm 10 \times 10^{-8}$, but which are not yet fully understood." Thus, there were possible systematic experimental uncertainties of the same order of magnitude as the expected effect. As Pickering states, these were novel experiments that used new and previously untried techniques. This tended to make the experimental results uncertain.

The theoretical calculations of the expected effect were also uncertain. The Oxford-Washington joint paper noted that Khriplovich had argued, in a soon-to-be-published paper, that the approximate theory overestimated R by a factor of approximately 1.5. In addition, the four calculations agreed with their mean only to within approximately ±25%. This made the largest and smallest calculated values of R differ by almost a factor of two.

In September 1977, both the Washington (Lewis et al.) and Oxford (Baird et al.) groups published more detailed accounts of their experiments, with somewhat revised results. Both groups again reported results in substantial disagreement with the predictions of the W-S theory, although the Washington group stated, "More complete calculations that include many-particle effects are clearly desirable." The Washington group reported a value of $R = (-0.7 \pm 3.2) \times 10^{-8}$, which was in disagreement with the prediction of approximately -25×10^{-8}. This value was also inconsistent with their earlier result of $(-8 \pm 3) \times 10^{-8}$. This inconsistency was not addressed in the published paper, but it was discussed within the atomic physics community and diminished the credibility of the result. The Oxford result was $R = (+2.7 \pm 4.7) \times 10^{-8}$, again in disagreement with the W-S prediction of approximately -30×10^{-8}. The researchers noted, however, that there was a systematic effect in their apparatus. They had found a change in ϕ_r, the rotation angle, caused by slight misalignment of the polarizers, optical rotation in the windows, and so forth, of order 20×10^{-8} radians. "Unfortunately, it varies with time over a period of minutes, and depends sensitively on the setting of the laser and the optical path through the polarizer. While we believe we understand this effect in terms of imperfections in the polarizers combined with changes in laser beam intensity distribution, we have been unable to reduce it significantly," acknowledged the experimenters. A systematic effect of this size certainly cast doubt on the result.

In the same issue of *Nature* in which the joint Oxford-Washington paper was published, Frank Close, a particle theorist, gave a summary that typified the opinion of the physics community (Close 1976): "Is parity violated in atomic physics? According to experiments being performed independently at Oxford and the University of Washington the answer may well be no. . . . This is a very interesting result in light of last month's report . . . claiming that parity is violated in high energy 'neutral current' interactions between neutrinos and matter." The experiment that Close referred to had concluded, "Measurements of R_ν and $R_{\bar{\nu}}$, the ratios of neutral current to charged current rates for ν and $\bar{\nu}$ [neutrino and antineutrino] cross sections, yield neutral current rates for ν and $\bar{\nu}$ that are consistent with a pure V-A interaction but 3 standard deviations from pure V or pure A, indicating the presence of parity nonconservation in the weak neutral current" (Benvenuti et al. 1976).

Close noted that as the atomic physics results stood, they appeared to be inconsistent with the predictions of the W-S model supplemented by atomic physics calculations. He also remarked, "At present the discrepancy can conceivably be the combined effect of systematic effects in atomic physics calculations and systematic uncertainties in the experiments." Pickering weighs in: "If one accepted the Washington-Oxford result, the obvious conclusion was that neutral current effects violated parity conservation in neutrino interactions and conserved parity in electron interactions" (1984b, 505–6). Close discussed this possibility along with another alternative that had an unexpected (on the basis of accepted theory) energy dependence, so that the high-energy experiments (the neutrino interactions) would be expected to show parity nonconservation and the low-energy atomic physics experiments would not. "Whether such a possibility could be incorporated into the unification ideas is not clear. It also isn't clear, yet, if we have to worry. However, the clear blue sky of summer now has a cloud in it. We wait to see if it heralds a storm" (quoted in Pickering 1984b, 506).

To Pickering, the 1977 publication of the Oxford and Washington results indicated that "the storm that Frank Close had glimpsed had materialised and was threatening to wash away the basic Weinberg-Salam model" (1984b, 298). There is some support for this view in D. Miller's summary of the Symposium on Lepton and Photon Interactions at High Energies, held in Hamburg August 25–31, 1977. Miller noted that Patrick Sandars had reported that neither his group at

Oxford nor the Washington group had seen any parity-violating effects and that "they have spent a great deal of time checking both their experimental sensitivity and the theory in order to be sure" (Miller 1977). Miller went on to state (as Pickering reported) that "S. Weinberg and others discussed the meaning of these results. It seems that the SU(2) × U(1) is to the weak interaction what the naive quark-parton model has been to QCD [quantum chromodynamics], a first approximation which has fitted a surprisingly large amount of data. Now it will be necessary to enlarge the model to accommodate the new quarks and leptons, the absence of atomic neutral currents, and perhaps also whatever it is that is causing trimuon events." (1984b, 298) Nevertheless, the uncertainty in these experimental results made the disagreement with the W-S theory only a worrisome situation and not a cause for epistemological crisis, as Pickering believes it should have been. In any event, the monopoly of Washington and Oxford was soon broken.

Discordant Results

Change came in 1978 when L. M. Barkov and M. S. Zolotorev, Soviet scientists from Novosibirsk, reported measurements on the same transition in bismuth studied by the Oxford group (Barkov and Zolotorev 1978a, 1978b). Their results agreed with the predictions of the W-S model. They gave a value for Ψ_{exp}/Ψ_{W-S} of (+1.4 ± 0.3) k, where Ψ was the angle of rotation of the plane of polarization by the bismuth vapor and k was a factor between 0.5 and 1.5, introduced because of inexact knowledge of the bismuth vapor. They concluded that their result did not contradict the predictions of the W-S model. Agreement with theoretical prediction depended (and still depends) on which method of calculation was chosen. A somewhat later paper changed the result to $\Psi_{exp}/\Psi_{W-S} = 1.1 ± 0.3$.

Subsequent papers reported more extensive data and found a value for R_{exp}/R_{theor} of 1.07 ± 0.14 (Barkov and Zolotorev 1979a, 1979b, 1980a, 1980b). They also reported that the latest unpublished results from the Washington and Oxford groups, communicated privately, now showed parity violation, although "the results of their new experiments have not reached good reproducibility" (1979a). These later results were also presented at a 1979 conference at which Dydak reviewed the situation.

According to Pickering, "The details of the Soviet experiment were not known to Western physicists, making a considered evaluation of its result problematic." This is incorrect. During September 1979, an international workshop devoted to neutral-current interactions in atoms was held in Cargese, France, and was attended by representatives of virtually all of the groups actively working in the field, including Oxford, Washington, and Novosibirsk. At that workshop, as Claude Bouchiat remarked in his workshop summary paper, "Professor Barkov, in his talk, gave a very detailed account of the Novosibirsk experiment and answered many questions concerning possible systematic errors" (1980, 364).

In early 1979, a Berkeley group reported atomic physics results for thallium that agreed with the predictions of the W-S model. Although these results were not definitive—they were only two standard deviations from zero—they did agree with the W-S theory in both sign and magnitude.

Thus, in mid-1979 the various atomic physics results concerning the W-S theory were inconclusive. The Oxford and Washington groups had originally reported a discrepancy, but their more recent results, although preliminary, showed the presence of the predicted parity-nonconserving effects. In addition, both the Soviet and Berkeley results agreed with the model. Dydak summarized the situation in a talk at a 1979 conference: "It is difficult to choose between the conflicting results in order to determine the eq [electron-quark] coupling constants. Tentatively, we go along with the positive results from Novosibirsk and Berkeley groups and hope that future development will justify this step (it cannot be justified at present, on clear-cut experimental grounds)." Pickering, however, reports, "Having decided not to take into account the Washington-Oxford results, Dydak concluded that parity violation in atomic physics was as predicted in the standard model" (1984b, 300).

I find no justification for Pickering's conclusion. Dydak was attempting to determine the best values for the parameters describing neutral-current electron scattering. Dydak had tentatively adopted the results in agreement with the W-S model, stipulating that experiments did not, at the time, justify a definite conclusion. He concluded nothing about the validity of the standard model. Bouchiat, in his 1979 summary paper, was more positive. After reviewing the Novosibirsk experiment as well as the conflict between the earlier and later Washington and Oxford results, he remarked, "*As a conclusion on this Bismuth session, one can say that parity violation has been observed roughly with the magnitude predicted by the Weinberg-Salam theory*," but even this more positive statement does not assert that the results agree precisely with the predictions of the theory, only that the experimental results were of the correct order of magnitude.

The situation became even more complex when a group at SLAC reported a result on the scattering of polarized electrons from deuterium that agreed with the W-S model (Prescott et al. 1978, 1979). This was the E122 experiment, which Pickering discusses. The researchers not only found the predicted scattering asymmetry but also obtained a value for $\sin^2\theta_W$ of 0.20 ± 0.03 (1978) and 0.224 ± 0.020 (1979), in agreement with other measurements ($\sin^2\theta_W$ is an important parameter in the W-S theory). They wrote, "We conclude that within experimental error our results are consistent with the W-S model, and furthermore our best value of $\sin^2\theta_W$ is in good agreement with the weighted average for the parameter obtained from neutrino experiments" (1979).

Pickering reviews the history of how the Washington-Oxford experiments, once accepted, came to be rejected and comes to these conclusions:

> In retrospect, it is easy to gloss the triumph of the standard model in the idiom of the "scientist's account": the Weinberg-Salam model, with an appropriate complement of quarks and leptons, made predictions which were verified by the facts. But missing from this gloss, as usual, is the element of choice. In assenting to the validity of the standard model, particle physicists chose to accept certain experimental reports and to reject others. . . . Between 1977 and 1979 there had been no *intrinsic* change in the status of the Washington-Oxford experiments. No data were withdrawn, and no fatal flaws in the experimental practice of either group had been proposed. What had changed was the context within which the data were assessed. Crucial to this change of context were the results of experiment E122 at SLAC. In its own way E122 was just as innovatory as the Washington-Oxford

experiments and its findings were, in principle, just as open to challenge. But particle physicists *chose* to accept the results of the SLAC experiment, *chose* to interpret them in terms of the standard model (rather than some alternative which might reconcile them with the atomic physics results) and therefore *chose* to regard the Washington-Oxford experiments as somehow defective in performance or interpretation. (1984b, 301)

Although I do not dispute Pickering's contention that choice was involved in the decision to accept the W-S model, I disagree with him about the reasons for that choice. Pickering maintains that social interests of competing scientific traditions were paramount; in the view of the physics community, eventually including atomic physicists, the choice was a reasonable one based on convincing experimental evidence.

Pickering seems to regard the original Oxford and Washington atomic physics results and the SLAC E122 experiment on the scattering of polarized electrons as having equal evidential weight. In fact, both the original experimental results on bismuth and the comparison between experiment and theory were quite uncertain. Add to this original uncertainty the later contradictory results of both the Washington and Oxford groups, the parity-nonconserving measurement of the Novosibirsk group, and the Berkeley result on thallium and the original results were, at the least, very uncertain.

In 1981, when they published their latest result "that agrees in sign and approximate magnitude with recent calculations based upon the Weinberg-Salam theory" (Hollister et al. 1981), the Washington collaborators conceded that their earlier results were unreliable. Although theoretical calculations had reduced the size of the expected effect by a factor of approximately three or four (depending on which method of calculation was used), the theoretical results still did not agree with the 1977 Washington-Oxford measurements. Furthermore, the 1981 Washington paper stated, "Our experiment and the bismuth optical-rotation experiments by three other groups [Oxford, Moscow, and Novosibirsk] have yielded results with significant mutual discrepancies far larger than the quoted errors." They also pointed out that their earlier measurements "were not mutually consistent."

The moral of this story is clear. These attempts to detect atomic-parity violation were extremely difficult experiments, beset with systematic errors of approximately the same size as the predicted effects. There is no reason to give priority to the earliest measurements, as Pickering does. It seems much more likely that these earlier results were less reliable because not all of the systematic errors were known.

The SLAC E122 Experiment and Its Support for the W-S Theory

Pickering is correct that the E122 experiment was innovative and that its findings were open to challenge. But Pickering neglects to emphasize that for this reason, the SLAC researchers presented a detailed discussion of their experimental apparatus and result and performed many checks on their experiment.

The experiment depended in large part on a new high-intensity source of polarized electrons. The polarization of the electron beam could be varied. "This reversal was done randomly on a pulse to pulse basis. The rapid reversals minimized the effects of drifts in the experiment, and the randomization avoided changing the helicity synchronously with periodic changes in experimental parameters" (Prescott et al. 1978). An earlier experiment had demonstrated that polarized electrons could be accelerated with negligible depolarization. In addition, both the sign and magnitude of the beam polarization were measured periodically by observing the known asymmetry in electron-electron scattering from a magnetized iron foil.

The experimenters also checked whether the apparatus produced spurious asymmetries. They measured the scattering by using the unpolarized beam from the regular SLAC electron gun, for which the measured asymmetry should be zero. They assigned polarizations to the unpolarized beam with the same random number generator that had determined the polarization of the electron beam used in the original experiment. They obtained a value for the experimental asymmetry that was consistent with zero and that also demonstrated that the apparatus could measure asymmetries of the order of 10^{-5}.

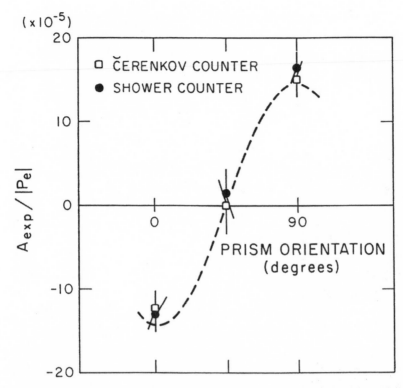

FIGURE 14.1. Experimental asymmetry as a function of prism angle for both the Cerenkov counter and the shower counter. The dashed line is the predicted behavior. From Prescott et al. (1978).

They also varied the polarization of the beam by changing the angle of a calcite prism. The results are shown in figure 14.1. Not only do the data fit the theoretically expected curve, but the fact that the results at 45° are consistent with zero indicates that other sources of error in the asymmetry are small. The graph shows the results for two different detectors, a nitrogen-filled Cerenkov counter and a lead glass shower counter. The consistency of the results increases belief in the validity of the measurements. The researchers wrote, "Although these two separate counters are not statistically independent, they were analyzed with independent electronics and respond quite differently to potential backgrounds. The consistency between these counters serves as a check that such backgrounds are small" (1978).

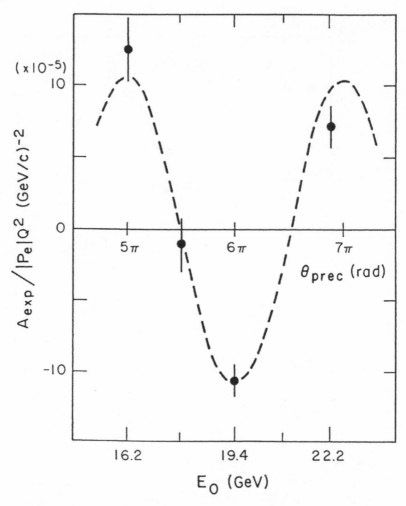

FIGURE 14.2. Experimental asymmetry as a function of beam energy. The expected behavior is the dashed line. From Prescott et al. (1978).

The electron beam polarization also depended on E_0, the beam energy. The expected distribution and the experimental data are shown in figure 14.2. The data clearly followed the theoretical predictions, and the fact that the value at 17.8 GeV (1 billion volts) was close to zero demonstrated that any disturbing effects were small.

A potential source of serious error came from small systematic differences in the beam parameters for the two polarizations. Small changes in beam position, angle, current, or energy could influence the measured yield and if correlated with reversals of beam polarization could cause apparent, but spurious, parity-violating asymmetries. These quantities were carefully monitored, and a feedback system was used to stabilize them: "Using the measured pulse to pulse beam information together with the measured sensitivities of the yield to each of the beam parameters, we made corrections to the asymmetries for helicity dependent differences in beam parameters. For these corrections, we have assigned a systematic error equal to the correction itself. The most significant imbalance was less than one part per million in E_0 [the beam energy] which contributed -0.26×10^{-5} to A/Q^2 [the experimental asymmetry]" (Prescott et al. 1978). This is to be compared to their final result of $A/Q^2 = (-9.5 \pm 1.6) \times 10^{-5}$ (GeV/c)$^{-2}$, which is obviously much larger. This was regarded by the physics community as a reliable and convincing result.

Pickering also claims that "hybrid models," which were able to accommodate the discordant results, were not discussed by the SLAC group. On the contrary, hybrid alternatives to the W-S theory were both considered and tested by E122. In their first paper, the researchers pointed out that the hybrid model was consistent with their data only for values of $\sin^2\theta_W < 0.1$, which was inconsistent with the measured value of approximately 0.23. In the second paper, they plotted their data as a function of $y = (E_0 - E')/E_0$, where E' is the energy of the scattered electron and E_0 is the electron-beam energy. Both models, W-S and the hybrid, made definite predictions for this graph. The results are shown in figure 14.3, and the superiority of the W-S model is obvious. For W-S they obtained the value $\sin^2\theta_W = 0.224 \pm 0.020$ with a probability of 40%. The hybrid model gave a value of 0.015 for $\sin^2\theta_W$ and a probability of 6×10^{-4}, which, they concluded, "appears to rule out this model."

The physics community chose to accept an experimental result that confirmed the W-S theory because it had been extremely carefully done and carefully checked. This position is supported by Bouchiat's 1979 summary. After hearing a detailed account of the SLAC experiment by Charles Prescott, he stated, "To our opinion, this experiment gave the first truly convincing evidence for parity violation in neutral current processes. . . . I would like to say that I have been very much

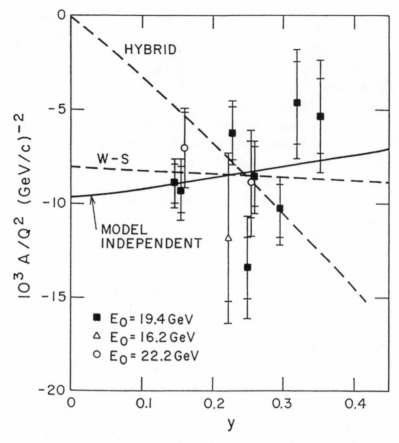

FIGURE 14.3. Asymmetries measured at three different energies plotted as a function of $y = (E_0 - E')/E_0$. The predictions of the hybrid model, the Weinberg-Salam theory, and a model independent calculation are shown. "The Weinberg-Salam model is an acceptable fit to the data; the hybrid model appears to be ruled out." From Prescott et al. (1978).

impressed by the care with which systematic errors have been treated in the experiment. It is certainly an example to be followed by all people working in this very difficult field" (Bouchiat 1980, 358). Bouchiat was a theoretical atomic physicist working on parity violation and not a member of the high-energy physics community. It was not the mighty traditions of high-energy physics that convinced him, but rather the experimental evidence. The physics community chose to await further developments in the atomic-parity-violating experiments because the results were uncertain.

Pickering is correct that no intrinsic change in the early Washington and Oxford results caused them to be rejected, but he fails to realize that they were never very convincing. They began as uncertain—although worrisome—and remained uncertain. The subsequent history shows that although other reliable atomic physics experiments confirm the W-S theory, the bismuth results, although generally in agreement with the predictions, are still somewhat uncertain. In addition, the most plausible alternative to the W-S model, which could reconcile the original atomic physics results with the electron-scattering data, was tested and found wanting. A choice certainly was made, but it was made on the basis of experimental evidence. The mutants died of natural causes, not from scientific opportunism.

The comparison between experiment and theory can often be extremely difficult. This is particularly true in this episode, conducted at the limit of what can be calculated confidently and measured reliably. In 1977, the atomic physics calculations of the expected parity-violating effects had a large uncertainty. Even today, estimates of the accuracy of the calculations are approximately 25%. The experimental results are also uncertain. (The smallest uncertainty in any experimental result is approximately 1%.) These experiments contain systematic errors, which may mimic or mask the expected result. Unlike statistical errors, which can be calculated precisely, systematic errors are extremely difficult both to detect and to estimate. The experimenters devoted considerable effort to finding and eliminating such systematic errors. Some of these errors have now been measured and corrected for. Others are still unknown. But physicists nonetheless have good reason to prefer the results from later, more sophisticated experiments.

Pickering remarks that by 1979 (and presumably to this day) there had been no intrinsic change in the early Washington and Oxford results. In the sense that no one knows with certainty why those early results were wrong, he is correct. However, since those early experiments, physicists have found new sources of systematic error that were not dealt with in the early experiments. The redesign of the apparatus has, in many cases, precluded testing whether or not these effects were significant in the older apparatus. Although it is not certain that these effects account for the earlier, presumably incorrect, results, there are reasonable grounds for believing that the later results are more accurate. The consistency of the later measurements, especially those done by different groups, enhances that belief.

It is clear that the relationship between theory and experiment is more complex than "the scientist proposes, Nature disposes." History shows that it is not at all clear at times just what the scientist is proposing. The theoretical predictions in this episode had considerable uncertainty. The uncertain and discordant experimental results show that one may not be able to see just how Nature is disposing. It seems clear, however, that a model of science that emphasizes evidence fits the history of this episode better than Pickering's model does. Scientists chose, on the basis of reliable experimental evidence provided by the SLAC E122 experiment, to accept the W-S theory. They chose to leave an apparent, but also quite uncertain, anomaly in the atomic-parity violation experiments for future investigation. Future work resolved the anomaly.

This episode demonstrates that scientists make judgments about the reliability of experimental results that coincide with epistemological evidence. The E122 experimenters argued for the validity of their experimental result with strategies that coincide with an epistemology of experiment. Therefore, the scientific community accepted their arguments.

Conclusion

NO EASY ANSWERS

The case studies presented in this book make a strong case that science is a reasonable enterprise that gives us knowledge of the natural world, based on valid experimental evidence and on reasoned and critical discussion. Although some of the hypotheses, suggestions, and experimental results turned out to be incorrect, these episodes are successes because the normal practice of science provided good reasons for the decisions that were made.

Even in the cases of scientific misbehavior—Rupp's fraudulent data on electron scattering and Millikan's cosmetic surgery on his oil-drop data—the issues were resolved reasonably. Rupp's results, which supported Mott's scattering theory, only added to the confusion concerning that theory in the 1930s. Most experimental results at the time disputed the theory. Subsequent experimental work revealed the reflection-transmission effect, and when experiments were designed to take that effect into account, the discord disappeared and the apparent refutation was succeeded by confirmation. Although Millikan's selectivity in both data and analysis procedures cannot be approved, it is clear that he knew that the effects of his selectivity in both data and calculational procedure were very small. The correctness of his result was established by the numerous repetitions of his important measurement of electron charge.

Another point that should be emphasized is that "wrong" physics is not bad physics. In the episodes of the nonconservation of parity, the fifth force in gravity, and the 17-keV (one thousand volts) neutrino, most physicists initially thought that these suggestions were incorrect. Sufficient grounds were found, however, to persuade a part of the physics community to investigate the suggestions further. Subsequent work provided good reasons to believe that parity was not conserved and that no fifth force or 17-keV neutrino existed. It could have turned out otherwise. That the last two suggestions turned out to be wrong does not mean that they should not have been proposed or further investigated. They could have been correct. Science would not progress if interesting but unlikely suggestions were not proposed and investigated.

The latter two cases had discordant experimental results. Similar discord marked the searches for gravity waves and the experiments on atomic-parity violation. The case studies show that the decisions regarding which results were correct were made on reasonable epistemological and methodological grounds.

To emphasize only difficulties and how they were overcome would be misleading. There are also examples of scientific success. One is the 100-year history of the neutrino, from the discovery of radioactivity to the recent Sudbury Neutrino Observatory studies on neutrino oscillations. It took 35 years for Wolfgang Pauli to suggest this elusive particle, another 25 years until Frederick Reines and Clyde Cowan experimentally demonstrated its existence, and 30 years from Raymond Davis's first report of the solar neutrino problem to its solution by neutrino oscillations. In that time scientists have learned a tremendous amount about a particle that interacts so weakly with matter that it is barely there. (Its interaction length in lead is 1,000 light-years.) It is now known that there are three types of neutrino and that they can transform into one another. Upper limits on the masses of the neutrinos have also been established, and the neutrino has been shown to possess handedness, like a screw. Unsolved problems remain—whether the neutrino and antineutrino are identical particles and what is the precise nature of the neutrino oscillations, for instance—but much has been learned.

The history of the neutrino emphasizes that there is very little instant rationality in science. In the case of the Konopinski-Uhlenbeck theory of β-decay, the best available experimental evidence initially suggested the need for an alternative to Enrico Fermi's theory. This was provided by Emil Konopinski and George Uhlenbeck's theory. Further experimental work seemed to confirm the latter. It was soon realized that the experimental results were incorrect and that an incorrect experiment-theory comparison was being made. When these problems were solved, the evidence favored Fermi's theory, as Konopinski himself stated. The process took 8 years. In the case of the fifth force, 4 years passed from the initial suggestion by Ephraim Fischbach and others to Orrin Fackler's comment, "The Fifth Force is dead." It was 8 years from John Simpson's first report of the 17-keV neutrino to the experiments at Tokyo and Argonne that showed it did not exist. One study is unfinished after 20 years. In 1982 Blas Cabrera published the results of his first search for magnetic monopoles. That search is continuing.

Even though it may take some time, the scientific community finds good reasons for its decisions.

Glossary

β-decay—A form of radioactivity in which an atomic nucleus transforms into another nucleus with the emission of an electron and a neutrino.

antiparticles—Our best theories, as well as considerable experimental evidence, indicate that every particle has an antiparticle. Particles and antiparticles have the same masses and lifetimes. In the case of charged particles, the antiparticle has a charge opposite to that of the particle. Thus, the positively charged positron is the antiparticle of the negatively charged electron. Electrically neutral particles such as the neutron and antineutron have different magnetic properties. Other neutral particles, such as the photon, the particle of light, and the neutral pion are their own antiparticles.

Balmer Series—Light is emitted by atoms. The frequencies or wavelengths of the light emitted by hydrogen form a simple numerical series called the Balmer series.

baryon—The atomic nucleus is composed of protons and neutrons, which are called baryons. The baryon number is the number of protons plus the number of neutrons. In any interaction the baryon number remains the same, or is conserved.

Bohr magneton—If we envision the electron as a small rotating charge, it behaves as if it were a small magnet. The strength of that magnet is the Bohr magneton.

boson—Elementary particles are divided into two classes; those with integral spin are called bosons, and those with half-integral spin are called fermions. They behave very differently.

centrifugation—The process in which a liquid is rotated at very high speed. The higher-density liquid is found further from the axis of rotation (see figure 2.2).

Cerenkov radiation—When an electrically charged particle travels faster than the speed of light in a material (it cannot travel faster than the speed of light in a vacuum), Cerenkov radiation is emitted. This radiation can be used to detect high-speed particles.

cloud chamber—A device for detecting elementary particles. It contains a supersaturated vapor. When an electrically charged particle passes through the vapor, it produces ions. Liquid droplets form along the ion trail, making the path of the particle visible.

Compton effect—The scattering of x-rays or γ-rays, high-energy electromagnetic radiation, by an electron.

DNA (deoxyribonucleic acid)—This is the molecule responsible for heredity. In 1953 Watson and Crick showed that it has a double-helical structure.

electron volt (eV, keV, MeV, etc.)—Physicists often measure energy in units of electron volts, the energy that an electron will acquire in passing through a potential difference of 1 V (1 keV = 1,000 V, 1 MeV = 1 million V, 1 GeV = 1 billion V).

electroweak interactions—Physicists usually speak of four interactions. In decreasing order of strength they are the strong, or nuclear, interaction, which holds the atomic nucleus together; the electromagnetic interaction, which holds atoms together; the weak interaction, responsible for radioactive decay; and the gravitational interaction. In the late 1960s, Abdus Salam, Steven Weinberg, and Sheldon Glashow unified the previously separate theories of the electromagnetic and the weak interactions into a single theory. More recently, the strong interactions have been added, forming what is now called the standard model.

excited state—Quantum mechanics teaches that atoms or nuclei can exist only in states with specific discrete energies. The state with the lowest energy is called the ground state. Higher-energy states are called excited states.

first-order correction—In considering perturbations to a system, it is possible to use only the largest perturbation and neglect the others because they are much smaller. This is called a first-order correction. The gravitational effect of Jupiter on the Earth's orbit is a first-order correction to the gravitational effect of the sun on the earth. (See also perturbation series expansion.)

ground state—Quantum mechanics teaches that atoms or nuclei can exist only in states with specific discrete energies. The state with the lowest energy is called the ground state.

hadrons—Particles that interact strongly, such as protons, neutrons, and prions.

half-life—The time it takes for one-half of the number of radioactive particles or nuclei to decay. On average after one half-life, one-half of the particles are left. After two half-lives, one-quarter of those initially present remain.

Heisenberg uncertainty principle—One of the interesting consequences of quantum mechanics and the best current theory of matter. It states that the position and momentum of an elementary particle cannot be measured simultaneously. It also states that an exact energy measurement cannot be made in a finite time.

inverse-square law—This law describes a force that varies inversely as the square of the distance between two objects. Doubling the distance between the objects reduces this force by a factor of four. Examples of such a law are the force of gravity and the force between two electrically charged particles.

isotopes—Atoms of a particular element that have different masses. The chemical behavior of isotopes of the same element is the same.

lepton—Originally a very light elementary particle with a spin of ½. It now includes the electron, the muon, the τ-lepton, their respective neutrinos, and the antiparticles of each.

light-year—the distance that light will travel in a year. It is approximately 9.46×10^{15} meters, or 5.88×10^{12} (million million) miles.

order of magnitude—A factor of 10. Thus 100 is an order of magnitude larger than 10, whereas 2 and 3 have the same order of magnitude.

perturbation series expansion—The use of a series of terms to replace a mathematical function. When a large physical effect is combined with several smaller effects, the smaller effects are described as perturbations. Thus in describing the orbits of the planets, the first term used is the gravitational interaction between the sun and the planet. The gravitational effects of the other planets, which are far smaller, are the perturbations. (See also first-order correction.)

phosphorescence—The delayed emission of light by a substance after it has been exposed to light.

photon—The fundamental particle of light, or electromagnetic radiation. Although light often behaves as if it were a wave, exhibiting interference phenomena, it also behaves as a particle in certain experiments.

piezo-electricity, or piezo-electric—When some substances are mechanically deformed, they produce a pulse of electricity. This is called piezo-electricity. A piezo-electric detector is attached to a gravity wave antenna to detect the deformation that should be produced if the antenna is struck by a gravity wave.

principle of equivalence—The principle that the effects of gravity and those of acceleration cannot be distinguished. It is a postulate of Albert Einstein's theory of general relativity. The principle requires that for any object, the gravitational mass, which is responsible for the force of gravity, and the inertial mass, a measure of how hard it is to accelerate an object, are equal.

quarks—The constituents of elementary particles; for example, a proton is believed to be made of three quarks and a meson of a quark and an antiquark. Quarks have an electrical charge that is either one-third or two-thirds that of the electron.

radioactivity—The transformation of an atomic nucleus into another atomic nucleus with the emission either of an electron and a neutrino or of an alpha particle, a positively charged helium nucleus. There is also a form of radioactivity in which the atomic nucleus goes from a higher-energy state to a lower-energy state with the emission of a γ-ray, high-energy electromagnetic radiation.

rise time—The time it takes for an electrical pulse to go from zero to its maximum value.

standard deviation—A measure of the statistical uncertainty in a given measurement. If an error is statistical, then the probability that the correct value is within one standard deviation of the average value is 68%.

statistical error or uncertainty—If a quantity is measured several times, the statistical error or uncertainty is a measure of the spread in the distribution of measurements.

systematic error or uncertainty—This is an error or uncertainty introduced by the method of measurement. For example, if human height were measured, and all the subjects stood on a 6-in box, this would cause in the measurement a systematic error of 6. It is usually quite difficult to estimate systematic errors in real experiments.

terminal velocity—If an object falls in a resistive medium such as air, it will eventually reach a highest velocity, or terminal velocity. This occurs when the resistance of the medium is equal to the force of gravity.

torque—The product of a force times its lever-arm distance, the perpendicular distance of the force from the axis of rotation. Exerting a force with a long wrench gives a larger torque than exerting the same force with a short wrench.

References

Abazov, A. I., O. L. Anosov, E. L. Faizov, et al. 1991. Search for neutrinos from the sun using the reaction $^{71}Ga(v_e,e^-)^{71}Ge$. *Physical Review Letters* 67:3332–35.

Ackermann, R. 1991. Allan Franklin, Right or wrong. In *PSA 1990*. Vol. 2, ed. A. Fine, M. Forbes, and L. Wessels, 451–57 East Lansing, MI: Philosophy of Science Association.

Adelberger, E. G., C. W. Stubbs, W. F. Rogers, et al. 1987. New constraints on composition-dependent interactions weaker than gravity. *Physical Review Letters* 59:849–52.

Ahmad, Q. R., R. C. Allen, T. C. Andersen, et al. 2001. Measurement of the rate of $v_e + d \rightarrow p + p + e^-$ interactions produced by ^8B solar neutrinos at the Sudbury Neutrino Observatory. *Physical Review Letters* 87:071301.

———. 2002. Direct evidence for neutrino flavor transformation from neutral-current interactions in the Sudbury Neutrino Observatory. *Physical Review Letters* 89:011301.

Altzitzoglou, T., F. Calaprice, M. Dewey, et al. 1985. Experimental search for a heavy neutrino in the beta spectrum of ^{35}S. *Physical Review Letters* 55:799–802.

Ander, M., M. A. Zumberge, T. Lautzenhiser, et al. 1989. Test of Newton's inverse-square law in the Greenland ice cap. *Physical Review Letters* 62:985–88.

Anderson, C. D. 1933. The positive electron. *Physical Review* 43:491–94.

Anselmann, P., W. Hampel, G. Heusser, et al. 1992a. Implications of the GALLEX determination of the solar neutrino flux. *Physics Letters* B285:390–97.

———. 1992b. Solar neutrinos observed by GALLEX at Gran Sasso. *Physics Letters* B285:376–89.

———. 1994. GALLEX results from the first 30 solar neutrino runs. *Physics Letters* B327:377–85.

Apalikov, A. M., S. D. Boris, A. I. Golutvin, et al. 1985. Search for heavy neutrinos in β decay. *JETP Letters* 42:289–93.

Aronson, S. H., G. J. Bock, H. Y. Cheng, et al. 1983a. Energy dependence of the fundamental parameters of the $K^0 - \overline{K}^0$ system. I. Experimental analysis. *Physical Review D* 28:476–94.

———— 1983b. Energy dependence of the fundamental parameters of the K^0 – $\overline{K}^0$ system. II. Theoretical formalism. *Physical Review D* 28:494–523.

Bahcall, J. N. 1964. Solar neutrinos I. Theoretical. *Physical Review Letters* 12:300–2.

Bahcall, J. N., N. A. Bahcall, and G. Shaviv. 1968. Present status of the theoretical predictions for the ^{37}Cl solar-neutrino experiment. *Physical Review Letters* 20:1209–12.

Bahcall, J. N., B. T. Cleveland, R. Davis, et al. 1978. A Proposed solar-neutrino experiment using ^{71}Ga. *Physical Review Letters* 40:1351–54.

Bahcall, J. N., and R. Davis 1989. An account of the development of the solar neutrino problem. In *Neutrino astrophysics*, ed. J. N. Bahcall, 487–530. Cambridge: Cambridge Univ. Press.

Bahcall, J. N., W. A. Fowler, I. Iben, et al. 1963. Solar neutrino flux. *Astrophysical Journal* 137:L344–L346.

Bahcall, J. N., and R. K. Ulrich. 1988. Solar models, neutrino experiments, and helioseismology. *Reviews of Modern Physics* 60:297–372.

Baird, P. E. G., M. W. S. Brimicombe, R. G. Hunt, et al. 1977. Search for parity-nonconserving optical rotation in atomic bismuth. *Physical Review Letters* 39:798–801.

Baird, P. E. G., M. W. S. Brimicombe, G. J. Roberts, et al. 1976. Search for parity non-conserving optical rotation in atomic bismuth. *Nature* 264:528–29.

Barkov, L. M., and M. S. Zolotorev. 1978a. Measurement of optical activity of bismuth vapor. *JETP Letters* 28:503–6.

————. 1978b. Observations of parity nonconservation in atomic transitions. *JETP Letters* 27:357–61.

————. 1979a. Parity violation in atomic bismuth. *Physics Letters* 85B:308–13.

————. 1979b. Parity violation in bismuth: experiment. In *International workshop on neutral current interactions in atoms in Cargese*, ed. W. L. Williams, 52–76. Washington, DC: National Science Foundation.

————. 1980a. Parity nonconservation in bismuth atoms and neutral weak-interaction currents. *JETP* 52:360–69.

————. 1980b. Parity violation in bismuth: experiment. In *International workshop on neutral current interactions in atoms (Cargese)*, ed. W. L. Williams, 52–76. Washington, DC: National Science Foundation.

Bartlett, D., S. Devons, and A. M. Sachs. 1962. Search for the decay mode: $\mu \to e + \gamma$. *Physical Review Letters* 8:120–23.

Bartlett, D. F., and W. L. Tew. 1989a. The fifth force: Terrain and pseudoterrain. In *Tests of fundamental laws in physics: Ninth Moriond Workshop*, ed. O. Fackler and J. Tran Thanh Van, 543–48. Gif sur Yvette, France: Editions Frontieres.

————. 1989b. Possible effect of the local terrain on the Australian fifth-force measurement. *Physical Review D* 40:673–75.

Becquerel, H. 1896a. Sur les propriétés différentes des radiations invisibles emises par les sels d'uranium, et du rayonnement de la paroi anticathodique d'un tube de crookes. *Comptes Rendus des Séances de L'Académie des Sciences* 122:762–67.

————. 1896b. Sur les radiations émises par phosphorescence. *Comptes Rendus des Séances de L'Académie des Sciences* 122:420–21.

————. 1896c. Sur les radiations invisibles émises par les corps phosphorescents. *Comptes Rendus des Séances de L'Académie des Sciences* 122:501–3.

————. 1896d. Sur les radiations invisibles émises par les sels d'uranium. *Comptes Rendus des Séances de L'Académie des Sciences* 122:689–94.

————. 1896e. Sur quelques propriétés nouvelles des radiations invisibles emises par divers corps phosphorescents. *Comptes Rendus des Séances de L'Académie des Sciences* 122:559–64.

Bennett, W. R. 1989. Modulated-source Eötvös experiment at Little Goose Lock. *Physical Review Letters* 62:365–68.

Benvenuti, A., D. Cline, F. Messing, et al. 1976. Evidence for parity nonconservation in the weak neutral current. *Physical Review Letters* 37:1039–42.

Bernstein, J. 1967. *A comprehensible world.* New York: Random House.

Bethe, H. A., and R. F. Bacher. 1936. Nuclear physics. *Reviews of Modern Physics* 8:82–229.

Bizzeti, P. G., A. M. Bizzeti-Sona, T. Fazzini, et al. 1988. New search for the 'fifth force' with the floating body method: Status of the Vallambrosa experiment. In *5th force neutrino physics: Eighth Moriond Workshop,* ed. O. Fackler and J. Tran Thanh Van, 501–14. Gif Sur Yvette, France: Editions Frontieres.

————. 1989. Search for a composition-dependent fifth force. *Physical Review Letters* 62:2901–4.

Bock, G. J., S. H. Aronson, K. Freudenreich, et al. 1979. Coherent K_S regeneration by protons from 30 to 130 GeV/c. *Physical Review Letters* 42:350–53.

Bohr, N. 1913. On the constitution of atoms and molecules. Part I. *Philosophical Magazine* 26:1–25.

————. 1929. β-Ray spectra and energy conservation. Unpublished manuscript. Niels Bohr Library, American Institute of Physics, New York.

————. 1932. Faraday Lecture: Chemistry and the quantum theory of atomic constitution. *Journal of the Chemical Society* 135:349–84.

Bohr, N., H. A. Kramers, and J. C. Slater. 1924. The quantum theory of radiation. *Philosophical Magazine* 47:785–802.

Bonvicini, G. 1993. Statistical issues in the 17-keV neutrino experiments. *Zeitschrift für Physik A* 345:97–117.

Bouchiat, C. 1980. Neutral current interactions in atoms. In *International workshop on neutral current interactions in atoms*, ed. W. L. Williams, 357–69. Washington, DC: National Science Foundation.

Boynton, P. 1990. New limits on the detection of a composition-dependent macroscopic force. In *New and exotic phenomena '90: Tenth Moriond Workshop*, ed. O. Fackler and J. Tran Thanh Van, 207–24. Gif sur Yvette, France: Editions Frontieres.

Bragg, W. H. 1904. On the absorption of alpha rays and on the classification of alpha rays from radium. *Philosophical Magazine* 8:719–25.

Brans, C., and R. H. Dicke. 1961. Mach's principle and a relativistic theory of gravitation. *Physical Review* 124:925–35.

Cabrera, B. 1982. First results from a superconductive detector for moving magnetic monopoles. *Physical Review Letters* 48:1378–81.

Cabrera, B., M. Taber, R. Gardner, et al. 1983. Upper limit on flux of cosmic-ray monopoles obtained with a three-loop superconductive detector. *Physical Review Letters* 51:1933–36.

Cairns, J. 1961. An estimate of the length of the DNA molecule of T2 Bacteriophage by autoradiography. *Journal of Molecular Biology* 3:756–61.

Chadwick, J. 1914. Intensitätsverteilung im magnetischen Spektrum der β-Strahlen von Radium B + C. *Verhandlungen der deutschen physikalischen Gesellschaft* 16:383–91.

Chase, C. 1929. A test for polarization in a beam of electrons by scattering. *Physical Review* 34:1069–74.

Chase, C. T. 1930a. The scattering of fast electrons by metals. I. *Physical Review* 36:984–87.

———. 1930b. The scattering of fast electrons by metals. II. *Physical Review* 36:1060–65.

Christenson, J. H., J. W. Cronin, V. L. Fitch, et al. 1964. Evidence for the 2π decay of the K^o_2 meson. *Physical Review Letters* 13:138–40.

Close, F. E. 1976. Parity violation in atoms? *Nature* 264:505–6.

Colella, R., A. W. Overhauser, and S. A. Werner. 1975. Observations of gravitationally induced quantum interference. *Physical Review Letters* 34:1472–74.

Collins, H. 1981. Stages in the empirical programme of relativism. *Social Studies of Science* 11:3–10.

———. 1985. *Changing order: Replication and induction in scientific practice.* London: Sage.

———. 1994. A strong confirmation of the experimenters' regress. *Studies in History and Philosophy of Modern Physics* 25:493–503.

Collins, H., and T. Pinch. 1993. *The golem: What everyone should know about science*. Cambridge: Cambridge University Press.

Cowan, C. L. 1964. Anatomy of an experiment: An account of the discovery of the neutrino. In *Annual Report of Regents of the Smithsonian Institution*, 409–30. Washington, DC: Smithsonian Institution.

Cowan, C. L., F. Reines, F. B. Harrison, et al. 1956. Detection of the free neutrino: A confirmation. *Science* 124:103–4.

Cowsik, R., N. Krishnan, S. N. Tandor, et al. 1988. Limit on the strength of intermediate-range forces coupling to isospin. *Physical Review Letters* 61:2179–81.

———. 1990. Strength of Intermediate-range forces: Coupling to isospin. *Physical Review Letters* 64:336–339.

Cox, R. T. 1973. Discovery story. *Adventures in Experimental Physics* gamma:149.

Cox, R. T., C. G. McIlwraith, and B. Kurrelmeyer. 1928. Apparent evidence of polarization in a beam of β-rays. *Proceedings of the National Academy of Sciences (USA)* 14:544–49.

Crane, H. R. 1948. The energy and momentum relations in the beta-decay, and the search for the neutrino. *Reviews of Modern Physics* 20:278–95.

Crookes, W. 1909. Antoine Henri Becquerel 1852–1908. *Proceedings of the Royal Society (London)* A83:xx–xxiii.

Danby, G., J.-M. Gaillard, K. Goulianos, et al. 1962. Observation of high-energy neutrino interactions and the existence of two kinds of neutrinos. *Physical Review Letters* 9:36–44.

Darwin, C. G. 1927. The electron as a vector wave. *Proceedings of the Royal Society (London)* A116:227–53.

Datar, V. M., C. V. K. Baba, S. K. Bhattacherjee, et al. 1985. Search for a heavy neutrino in the β-decay of ^{35}S. *Nature* 318:547–8.

Davis, R. 1964. Solar neutrinos II. Experimental. *Physical Review Letters* 12:303–5.

Davis, R., D. S. Harmer, and K. C. Hoffman. 1968. Search for neutrinos from the sun. *Physical Review Letters* 20:1205–9.

Davisson, C., and L. H. Germer. 1927. Diffraction of electrons by a crystal of nickel. *Physical Review* 30:705–40.

De Broglie, L. 1923. Ondes et quanta. *Comptes Rendus des Séances de L'Académie des Sciences* 177:507–10.

Decamp, D., B. Deschizeaux, C. Goy, et al. 1990. Measurement of electroweak parameters from Z decays into fermion pairs. *Zeitschrift für Physics C* 48:365–91.

Delbruck, M., and G. S. Stent. 1957. On the mechanism of DNA replication. In *The chemical basis of heredity*, ed. W. D. McElroy and B. Glass, 699–736. Baltimore: Johns Hopkins Press.

De Rujula, A. 1986. Are there more than four? *Nature* 323:760–61.

Dirac, P. A. M. 1929. Letter to Bohr, November 26.

————. 1931. Quantified singularieties in the electromagnetic field. *Proceedings of the Royal Society (London)* 133:60–72.

————. 1948. The theory of magnetic poles. *Physical Review* 74:817–30.

Dydak, F. 1979. Neutral currents. In *Proceeding of the Conference on High Energy Physics*, 25–49. Geneva: CERN.

Eckhardt, D. 1988. Results of a tower gravity experiment. In *5th Force neutrino physics: Eighth Moriond Workshop*, ed. O. Fackler and J. Tran Thanh Van, 577–83. Gif sur Yvette, France: Editions Frontieres.

Eckhardt, D., C. Jekeli, A. R. Lazarewicz, et al. 1989. Evidence for non-Newtonian gravity: Status of the AFGL experiment January 1989. In *Tests of fundamental laws in physics: Ninth Moriond Workshop*, ed. O. Fackler and J. Tran Thanh Van, 525–27. Gif sur Yvette, France: Editions Frontieres.

Eddington, A. S. 1927. *Stars and atoms.* New Haven, CT: Yale University Press.

Einstein, A. 1907. Relativitätsprinzip und die aus demselben gezogenen Folgerungen. *Jahrbuch Radioaktivität* 4:411–62.

Ellis, C. D., and W. J. Henderson. 1934. Artificial radioactivity. *Proceedings of the Royal Society (London)* A146:206–16.

Ellis, C. D., and W. A. Wooster. 1925. The β-ray type of disintegration. *Proceedings of the Cambridge Philosophical Society* 22:849–60.

————. 1927. The average energy of disintegration of radium E. *Proceedings of the Royal Society (London)* A117:109–23.

Eötvös, R., D. Pekar, and E. Fekete. 1922. Beiträge zum Gesetze der Proportionalität von Trägheit und Gravität. *Annalen der Physik (Leipzig)* 68:11–66.

Fairbank, W. M. 1988. Summary talk on fifth force papers. In *5th Force neutrino physics: Eighth Moriond Workshop*, ed. O. Fackler and J. Tran Thanh Van, 629–44. Gif sur Yvette, France: Editions Frontieres.

Fairbank Jr., W. M., and A. Franklin. 1982. Did Millikan observe fractional charges on oil drops? *American Journal of Physics* 50:394–97.

Fermi, E. 1934a. Attempt at a theory of β-rays. *Il Nuovo Cimento* 11:1–21.

————. 1934b. Versuch einer Theorie der β-Strahlen. *Zeitschrift für Physik* 88:161–77.

————. 1950. *Nuclear physics.* Chicago: University of Chicago Press.

Feynman, R. P. 1986. Letter to the editor. Los Angeles Times, January 23.

Feynman, R. P., R. B. Leighton, and M. Sands. 1963. *The Feynman lectures on physics*. Reading, MA: Addison-Wesley Publishing Company.

Fischbach, E. 1980. Tests of general relativity at the quantum level. In *Cosmology and gravitation*, ed. P. Bergmann and V. De Sabbata, 359–73. New York: Plenum.

Fischbach, E., S. Aronson, C. Talmadge, et al. 1986a. Reanalysis of the Eötvös experiment. *Physical Review Letters* 56:3–6.

———. 1986b. Response to Thodberg. *Physical Review Letters* 56:2424.

Fitch, V. L., M. V. Isaila, and M. A. Palmer. 1988. Limits on the existence of a material-dependent intermediate-range force. *Physical Review Letters* 60:1801–4.

Ford, E. B. 1937. Problems of heredity in the Lepidoptera. *Biological Reviews* 12:461–503.

———. 1940. Genetic research on the Lepidoptera. *Annals of Eugenics* 10:227–52.

Frankel, S., J. Halpern, L. Holloway, et al. 1962. New limit on the e + γ decay mode of the muon. *Physical Review Letters* 8:123–25.

Franklin, A. 1981. Millikan's published and unpublished data on oil drops. *Historical Studies in the Physical Sciences* 11:185–201.

———. 1986. *The neglect of experiment*. Cambridge: Cambridge University Press.

———. 1990. *Experiment, right or wrong*. Cambridge: Cambridge University Press.

———. 1991. Do mutants have to be slain, or do they die of natural causes. In *PSA 1990*. Vol. 2, ed. A. Fine, M. Forbes, and L. Wessels, 487–94. East Lansing, MI: Philosophy of Science Association.

———. 1993a. Discovery, pursuit, and justification. *Perspectives on Science* 1:252–84.

———. 1993b. *The rise and fall of the fifth force: Discovery, pursuit, and justification in modern physics*. New York: American Institute of Physics.

———. 1994. How to avoid the experimenters' regress. *Studies in History and Philosophy of Modern Physics* 25:97–121.

———. 1995a. The appearance and disappearance of the 17-keV neutrino. *Reviews of Modern Physics* 67:457–90.

———. 1995b. The resolution of discordant results. *Perspectives on Science* 3:346–420.

———. 1997a. Are there really electrons? Experiment and reality. *Physics Today* 50:26–33.

———. 1997b. Calibration. *Perspectives on Science* 5:31–80.

———. 2000. *Are there really neutrinos? An evidential history.* Cambridge, MA: Perseus.

Frauenfelder, H., and E. M. Henley. 1975. *Nuclear and particle physics.* Reading, MA: W. A. Benjamin.

French, A. P. 1999. The strange case of Emil Rupp. *Physics in Perspective* 1:3–21.

Friedman, J. L., and V. L. Telegdi. 1957. Nuclear emulsion evidence for parity nonconservation in the decay chain pi - mu-e. *Physical Review* 105:1681–82.

Fujii, Y. 1971. Dilatonal possible non-Newtonian gravity. *Nature* 234:5–7.

———. 1972. Scale invariance and gravity of hadrons. *Annals of Physics (NY)* 69:494–521.

———. 1974. Scalar-tensor theory of gravitation and spontaneous breakdown of scale invariance. *Physical Review D* 9:874–76.

Fukuda, Y., T. Hayakawa, E. Ichihara, et al. 1998. Evidence for oscillation of atmospheric neutrinos. *Physical Review Letters* 81:1562–67.

Fukuda, Y., T. Hayakawa, K. Inoue, et al. 1994. Atmospheric v_μ/v_e ratio in the multi-GeV energy range. *Physics Letters* B335:237–45.

Gardner, R. D., B. Cabrera, M. E. Huber, et al. 1991. Search for cosmic-ray monopoles using a three-loop superconductive detector. *Physical Review D* 44:622–35.

Garwin, R. L. 1974. Detection of gravity waves challenged. *Physics Today* 27:9–11.

Garwin, R. L., L. M. Lederman, and M. Weinrich. 1957. Observation of the failure of conservation of parity and charge conjugation in meson decays: The magnetic moment of the free muon. *Physical Review* 105:1415–17.

Gerlach, W., and O. Stern. 1921. Der experimentelle Nachweis des magnetischen Moments des Silberatoms. *Zeitschrift für Physik* 8:110–11.

———. 1922. Der experimentelle Nachweis der Richtungsquantelung. *Zeitschrift für Physik* 9:349–52.

Gibbons, G. W., and B. F. Whiting. 1981. Newtonian gravity measurements impose constraints on unification theories. *Nature* 291:636–38.

Glashow, S. L. 1991. A Novel neutrino mass hierarchy. *Physics Letters* 256B:255–57.

Gribov, V., and B. Pontecorvo. 1969. Neutrino astronomy and lepton charge. *Physics Letters* 28B:493–96.

Grodzins, L. 1959. The history of double scattering of electrons and evidence for the polarization of beta rays. *Proceedings of the National Academy of Sciences (USA)* 45:399–405.

———. 1973. Early experiments on parity nonconservation. *Adventures in Experimental Physics* gamma:154–60.

Gunns, A. F., and J. R. Goodstein. 1975. *Guide to the Robert Andrews Millikan collection*. New York: American Institute of Physics.

Hacking, I. 1981. Do we see through a microscope. *Pacific Philosophical Quarterly* 63:305–22.

———. 1983. *Representing and Intervening*. Cambridge: Cambridge University Press.

Hagiwara, K., K. Hikasa, K. Nakamura, et al. 2002. Review of particle physics. *Physical Review D* 66:1–974

Hahn, O. 1966. *Otto Hahn: A Scientific Autobiography*. New York: Charles Scribner's Sons.

Hahn, O., and L. Meitner. 1908a. Über die Absorption der β-Strahlen einiger Radioelemente. *Physikalische Zeitschrift* 9:321–33.

———. 1908b. Über die β-Strahlen des Aktiniums. *Physikalische Zeitschrift* 9:697–704.

———. 1909a. Über das Absorptionsgesetz der β-Strahlen. *Physikalische Zeitschrift* 10:948–50.

———. 1909b. Über eine typische β-Strahlung des eigentlicher Radiums. *Physikalische Zeitschrift* 10:741–45.

———. 1910. Eine neue β-Strahlung beim Thorium X; Analogien in der Uran- und Thoriumreihe. *Physikalische Zeitschrift* 11:493–97.

Heckel, B. R., E. G. Adelberger, C. W. Stubbs, et al. 1989. Experimental bounds on interactions mediated by ultralow-mass bosons. *Physical Review Letters* 63:2705–8.

Hetherington, D. W., R. L. Graham, M. A. Lone, et al. 1987. Upper limits on the mixing of heavy neutrinos in the beta decay of ^{63} Ni. *Physical Review C* 36:1504–13.

Hime, A. 1992. Pursuing the 17 keV neutrino. *Modern Physics Letters A* 7:1301–14.

———. 1993. Do scattering effects resolve the 17-keV conundrum? *Physics Letters* 299B:165–73.

Hime, A., and N. A. Jelley. 1991. New evidence for the 17 keV neutrino. *Physics Letters* 257B:441–49.

Hime, A., and J. J. Simpson. 1989. Evidence of the 17-keV neutrino in the β spectrum of ^{3}H. *Physical Review D* 39:1837–50.

Hirata, K. S., K. Inoue, T. Ishida, et al. 1991. Real-time directional measurement of ^{8}B solar neutrinos in the Kamiokonde II detector. *Physical Review D* 44:2241–60.

———. 1992. Observation of a small atmospheric v_μ/v_e ratio in Kamiokonde. *Physics Letters* B280:146–52.

Hirata, K. S., K. Inoue, T. Kajita, et al. 1990. Constraints on neutrino-oscillation parameters from the Kamiokonde-II solar-neutrino data. *Physical Review Letters* 65:1301–4.

Hirata, K. S., T. Kajita, M. Koshiba, et al. 1988. Experimental study of the atmospheric neutrino flux. *Physics Letters* B205:416–20.

Hollister, J. H., G. R. Apperson, L. L. Lewis, et al. 1981. Measurement of parity nonconservation in atomic bismuth. *Physical Review Letters* 46:643–46.

Holmes, F. L. 2001. *Meselson, Stahl, and the replication of DNA, a history of the most beautiful experiment in biology.* New Haven, CT: Yale University Press.

Holton, G. 1978. Subelectrons, presuppositions, and the Millikan-Ehrenhaft debate. *Historical Studies in the Physical Sciences* 9:166–224.

Huber, M. E., B. Cabrera, M. A. Taber, et al. 1990. Limit on the flux of cosmic-ray monopoles from operations of an eight-loop superconducting detector. *Physical Review Letters* 64:835–38.

———. 1991. Search for a flux of cosmic-ray monopoles with an eight-channel superconducting detector. *Physical Review D* 44:636–60.

Hulse, R. A., and J. H. Taylor. 1975. A deep sample of new pulsars and their spatial extent in the galaxy. *Astrophysical Journal* 201:L55-L59.

Jekeli, C., D. H. Eckhardt, and A. J. Romaides. 1990. Tower gravity experiment: No evidence for non-Newtonian gravity. *Physical Review Letters* 64:1204–6.

Jordan, P., and R. d. L. Kronig. 1927. Movement of the lower jaw of cattle during mastication. *Nature* 120:807.

Kammeraad, J., P. Kasameyer, O. Fackler, et al. 1990. New results from Nevada: A test of Newton's law using the BREN tower and a high density gravity survey. In *New and exotic phenomena '90: Tenth Moriond Workshop*, ed. O. Fackler and J. Tran Thanh Van, 245–54. Gif sur Yvette, France: Editions Frontieres.

Kasameyer, P., J. Thomas, O. Fackler, et al. 1989. A test of Newton's law of gravity using the BREN tower, Nevada. In *Tests of fundamental laws in physics: Ninth Moriond Workshop*, ed. O. Fackler and J. Tran Thanh Van, 529–42. Gif sur Yvette, France: Editions Frontieres.

Kaufmann, W. 1902. Die elektromagnetische Masse des Elektrons. *Physikalische Zeitschrift* 4:54–7.

Kawakami, H., S. Kato, T. Ohshima, et al. 1992. High sensitivity search for a 17 keV neutrino. Negative indication with an upper limit of 0.095%. *Physics Letters* 287B:45–50.

Kelvin. 1897. Contact electricity and electrolysis according to Father Boscovich. *Nature* 56:84–85.

Kettlewell, H. B. D. 1955. Selection experiments on industrial melanism in the Lepidoptera. *Heredity* 9:323–42.

————. 1956. Further selection experiments on industrial melanism in the Lepidoptera. *Heredity* 10:287–301.

————. 1958. A survey of the frequencies of *Biston betularia* (L.) (Lep.) and its melanic forms in Great Britain. *Heredity* 12:51–72.

Konopinski, E. 1943. Beta-decay. *Reviews of Modern Physics* 15:209–45.

Konopinski, E., and G. Uhlenbeck. 1935. On the Fermi theory of radioactivity. *Physical Review* 48:7–12.

————. 1941. On the theory of β-radioactivity. *Physical Review* 60:308–20.

Kurie, F. N. D., J. R. Richardson, and H. C. Paxton. 1936. The radiations from artificially produced radioactive substances. *Physical Review* 49:368–81.

Kuroda, K., and N. Mio. 1989a. Galilean test for composition-dependent force. In *Proceedings of the Fifth Marcel Grossman Conference on General Relativity*, ed. D. G. Blair and M. J. Buckingham, 1569–72. Singapore: World Scientific.

————. 1989b. Test of a composition-dependent force by a free-fall interferometer. *Physical Review Letters* 62:1941–44.

Kuz'min, V. A. 1966. Detection of solar neutrinos by means of the $Ga^{71}(v,e^-)Ge^{71}$reaction. *JETP* 22:1051–52.

Lawson, J. L. 1939. The Beta-ray spectra of phosphorus, sodium, and cobalt. *Physical Review* 56:131–36.

Lawson, J. L., and J. M. Cork. 1940. The radioactive isotopes of indium. *Physical Review* 57:982–94.

Lee, T. D. 1971. The history of weak interactions. Speech at the Columbia University Physics Department, March 26.

Lee, T. D., and C. N. Yang. 1956. Question of parity nonconservation in weak interactions. *Physical Review* 104:254–58.

Lehninger, A. L. 1975. *Biochemistry*. New York: Worth.

Levine, G., ed. 1987. *One culture: Essays in science and literature*. Madison: The University of Wisconsin Press.

Levine, J. L. 2004. Early gravity-wave detection experiments, 1960–1975. *Physics in Perspective* 6:42–75.

Lewis, L. L., J. H. Hollister, D. C. Soreide, et al. 1977. Upper limit on parity-nonconserving optical rotation in atomic bismuth. *Physical Review Letters* 39:795–98.

Lindhard, J., and P. G. Hansen. 1986. Atomic effects in low-energy beta decay: The case of tritium. *Physical Review Letters* 57:965–67.

Livingston, M. S., and H. A. Bethe. 1937. Nuclear physics. *Reviews of Modern Physics* 9:245–390.

Lusignoli, M., and A. Pugliese. 1986. Hyperphotons and K-meson decays. *Physics Letters* 171B:468–70.

Lynch, M. 1991. Allan Franklin's transcendental physics. In *PSA 1990*. Vol. 2, ed. A. Fine, M. Forbes, and L. Wessels, 471–85. East Lansing, MI: Philosophy of Science Association.

Maddox, J. 1986. Newtonian gravity corrected. *Nature* 319:173.

Markey, H., and F. Boehm. 1985. Search for admixture of heavy neutrinos with masses between 5 and 55 keV. *Physical Review C* 32:2215–16.

Meitner, L. 1922a. Über den Zusammenhang zwischen β- und γ-Strahlen. *Zeitschrift für Physik* 9:145–52.

———. 1922b. Über die Entstehung der β-Strahl-Spektren radioaktiver Substanzen. *Zeitschrift für Physik* 9:131–44.

Meitner, L., and W. Orthmann. 1930. Über eine absolute Bestimmung der Energie der primären β-Strahlen von Radium E. *Zeitschrift für Physik* 60:143–55.

Meselson, M., and F. W. Stahl. 1958. The replication of DNA in Escherichia coli. *Proceedings of the National Academy of Sciences (USA)* 44:671–82.

Mikheyev, S. P., and A. Y. Smirnov. 1985. Resonance enhancement of oscillations and solar neutrino spectroscopy. *Soviet Journal of Nuclear Physics* 42:913–17.

———. 1986. Resonant amplification of ν oscillations in matter and solar-neutrino spectroscopy. *Il Nuovo Cimento* 9C:17–26.

Mikkelsen, D. R., and M. J. Newman. 1977. Constraints on the gravitational constant at large distances. *Physical Review D* 16:919–26.

Miller, A. 1981. *Albert Einstein's special theory of relativity*. Reading, MA: Addison-Wesley.

Miller, D. J. 1977. Elementary particles—a rich harvest. *Nature* 269:286–88.

Millikan, R. A. 1911. The isolation of an ion, a precision measurement of its charge, and the correction of Stokes's law. *Physical Review* 32:349–97.

———. 1913. On the elementary electrical charge and the Avogadro constant. *Physical Review* 2:109–43.

———. 1917. *The electron*. Chicago: University of Chicago Press.

Morrison, D. 1992. Review of 17 keV neutrino experiments. In *Joint International lepton-photon symposium and europhysics conference on high energy physics*. Vol. 1, ed. S. Hegarty, K. Potter, and E. Quercigh, 599–605. Geneva, Switzerland: World Scientific.

Mortara, J. L., I. Ahmad, K. P. Coulter, et al. 1993. Evidence against a 17 keV neutrino from ³⁵S beta decay. *Physical Review Letters* 70:394–97.

Mott, N. F. 1929. Scattering of fast electrons by atomic nuclei. *Proceedings of the Royal Society (London)* A124:425–42.

Nelson, P. G., D. M. Graham, and R. D. Newman. 1990. Search for an intermediate-range composition-dependent force coupling to N-Z. *Physical Review D* 42:963–76.

Newman, R., D. Graham, and P. Nelson. 1989. A "fifth force" search for differential accleration of lead and copper toward lead. In *Tests of fundamental laws in physics: Ninth Moriond Workshop*, ed. O. Fackler and J. Tran Thanh Van, 459–72. Gif sur Yvette, France: Editions Frontieres.

Niebauer, T. M., M. P. McHugh, and J. E. Faller. 1987. Galilean test for the fifth force. *Physical Review Letters* 59:609–12.

O'Conor, J. S. 1937. The beta-ray spectrum of radium E. *Physical Review* 52:303–14.

Ohi, T., M. Nakajima, H. Tamura, et al. 1985. Search for heavy neutrinos in the beta decay of ^{35}S. Evidence against the 17 keV heavy neutrino. *Physics Letters* 160B:322–24.

Ohshima, T. 1993. 0.073% (95% CL) upper limit on 17 keV neutrino admixture. In *XXVI International Conference on High Energy Physics*. Vol. 1, ed. J. R. Sanford, 1128–35. Dallas: American Institute of Physics.

Ohshima, T., H. Sakamoto, T. Sato, et al. 1993. No 17 keV neutrino: Admixture < 0.073% (95% C.L.). *Physical Review D* 47:4840–56.

Pais, A. 1986. *Inward bound*. New York: Oxford University Press.

Parker, R. L., and M. A. Zumberge. 1989. An analysis of geophysical experiments to test Newton's law of gravity. *Nature* 342:29–32.

Pauli, W. 1922. Letter to Gerlach.

———. 1929. Letter to Bohr.

———. 1933. Die Allgemeinen Prinzipen der Wellenmechanik. *Handbuch der Physik* 24:83–272.

Paxton, H. C. 1937. The radiations from artificially produced radioactive substances. III. Details of the beta-ray spectrum of P^{32}. *Physical Review* 51:170–77.

Perl, M. L., G. S. Abrams, A. M. Boyarski, et al. 1975. Evidence for anomalous lepton production in e^{+}-e^{-} annihilation. *Physical Review Letters* 35:1489–92.

Perl, M. L., G. J. Feldman, G. S. Abrams, et al. 1976. Properties of anomalous eμ events produced in e^{+}e^{-} annihilation. *Physics Letters* 63B:466–70.

———. 1977. Properties of the proposed τ charged lepton. *Physics Letters* 70B:487–90.

Pickering, A. 1984a. Against putting the phenomena first: The discovery of the weak neutral current. *Studies in the History and Philosophy of Science* 15:85–117.

———. 1984b. *Constructing quarks*. Chicago: University of Chicago Press.

———. 1991. Reason enough? More on parity violation experiments and electroweak gauge theory. In *PSA 1990*. Vol. 2, ed. A. Fine, M. Forbes, and L. Wessels, 459–69. East Lansing, MI: Philosophy of Science Association.

Piilonen, L., and A. Abashian. 1992. On the strength of the evidence for the 17 keV neutrino. In *Progress in atomic physics, neutrinos and gravitation: Proceedings of the 27th Moriond Workshop*, ed. O. Fackler and J. Tran Thanh Van, 225–42. Gif Sur Yvette, France: Editions Frontieres.

Pontecorvo, B. 1946. Inverse β process. In Neutrino Physics, ed. K. Winter, 25–31. Cambridge: Cambridge University Press.

———. 1958. Mesonium and antimesonium. *Soviet Physics JETP* 6:429–31.

———. 1960. Electron and muon neutrinos. *Soviet Physics JETP* 10:1236–40.

———. 1968. Neutrino experiments and the problem of conservation of leptonic charge. *Soviet Physics JETP* 26:984–88.

Prescott, C. Y., W. B. Atwood, R. L. A. Cottrell, et al. 1978. Parity non-conservation in inelastic electron scattering. *Physics Letters* 77B:347–52.

———. 1979. Further measurements of parity non-conservation in inelastic electron scattering. *Physics Letters* 84B:524–28.

Raab, F. J. 1987. Search for an intermediate-range interaction: Results of the Eot-Wash I experiment. In *New and exotic phenomena: Seventh Moriond Workshop*, ed. O. Fackler and J. Tran Thanh Van, 567–77. Les Arcs, France: Editions Frontieres.

Reines, F. 1982a. Fifty years of neutrino physics: Early experiments. In *Neutrino physics and astrophysics*, ed. E. Fiorini, 11–28. New York: Plenum.

———. 1982b. Neutrinos to 1960—personal recollections. *Journal de Physique* 43, suppl. C8:237–60.

Reines, F., C. L. Cowan, F. B. Harrison, et al. 1960. Detection of the free antineutrino. *Physical Review* 117:159–73.

Richardson, O. W. 1934. The low energy β-rays of radium E. *Proceedings of the Royal Society (London)* A147:442–54.

Richter, H. 1937. Zweimalige Streuung schneller Elektronen. *Annalen der Physik* 28:553–54.

Roehrig, J., A. Gsponer, W. R. Molzon, et al. 1977. Coherent regeneration of K_s's by carbon as a test of regge-pole-exchange theory. *Physical Review Letters* 38:1116–19.

Rose, M. E., and H. A. Bethe. 1939. On the absence of polarization in electron scattering. *Physical Review* 55:277–89.

Rowley, J. K., B. T. Cleveland, and R. Davis. 1985. The chlorine solar neutrino experiment. In *Solar neutrinos and neutrino astronomy*, ed. M. L. Cherry, K. Lande, and W. A. Fowler, 1–21. New York: American Institute of Physics.

Rudge, D. W. 1998. A Bayesian analysis of strategies in evolutionary biology. *Perspectives on Science* 6:341–60.

————. 2001. Kettlewell from an error statistician's point of view. *Perspectives on Science* 9:59–77.

Rupp, E. 1929. Versuche zur Frage nach einer Polarisation der Elektronenwelle. *Zeitschrift für Physik* 53:548–52.

————. 1930. Über eine unsymmetrische Winkelverteilung zweifach reflektierter Elektronen. *Zeitschrift für Physik* 61:158–9.

————. 1931. Direkte Photographie der Ionisierung in Isolierstoffen. *Naturwissenschaften* 19:109.

————. 1932a. Neure Versuche zur Polarisation der Elektronen. *Physikalische Zeitschrift* 33:937–40.

————.1932b. Über die Polarisation der Elektronen bei zweimaliger 90°— Streuung. *Zeitschrift für Physik* 79:642–54.

————. 1932c. Versuche zum Nachweis einer Polarisation der Elektronen. *Physikalische Zeitschrift* 33:158–64.

————. 1934. Polarisation der Elektronen an freien Atomen. *Zeitschrift für Physik* 88:242–46

————. 1935. Mitteilung. *Zeitschrift für Physik* 95:810.

Rupp, E., and L. Szilard. 1931. Beeinflussung 'polarisierter' Elektronenstrahlen durch Magnetfelder. *Naturwissenschaften* 19:422–23.

Rutherford, E. 1899. Uranium radiation and the electrical conduction produced by it. *Philosophical Magazine* 47:109–63.

————. 1913. *Radioactive substances and their radiations.* Cambridge: Cambridge University Press.

————. 1929. Letter to Bohr, November 19.

Rutherford, E., and H. Robinson. 1913. The analysis of the β rays from radium B and radium C. *Philosophical Magazine* 26:717–29.

Salpeter, E. E. 1968. Neutrinos from the sun. *Comments on Nuclear and Particle Physics* 2:97–102.

Sargent, B. W. 1932. Energy distribution curves of the disintegration electrons. *Proceedings of the Cambridge Philosophical Society* 24:538–53.

————. 1933. The maximum energy of the β-rays from uranium X and other bodies. *Proceedings of the Royal Society (London)* A139:659–73.

Schmidt, H. W. 1906. Über die Absorption der β-Strahlen des Radiums. *Physikalische Zeitschrift* 7:764–66.

————. 1907. Einige Versuche mit β-Strahlen von Radium E. *Physikalische Zeitschrift* 8:361–73.

Schwartz, M. 1960. Feasibility of using high-energy neutrinos to study the weak interactions. *Physical Review Letters* 4:306–7.

Schwarzschild, B. 1986. Reanalysis of old Eötvös data suggests 5th force . . . to some. *Physics Today* 39:17–20.

Segre, E. 1980. *From X-rays to quarks*. Berkeley: University of California Press.

Shaviv, G., and J. Rosen, eds. 1975. *General relativity and gravitation: Proceedings of the Seventh International Conference (GR7), Tel-Aviv University, June 23–28, 1974*. New York: John Wiley.

Sime, R. L. 1996. Lise Meitner, a life in physics. Berkeley: University of California Press.

Simpson, J. J. 1985. Evidence of heavy-neutrino emission in beta decay. *Physical Review Letters* 54:1891–93.

——. 1986a. Evidence for a 17-keV neutrino in ^{3}H and ^{35}S β spectra. In *'86 Massive neutrinos in astrophysics and in particle physics: Proceedings of the Sixth Moriond Workshop*, ed. O. Fackler and J. Tran Thanh Van, 565–77. Gif sur Yvette, France: Editions Frontieres.

——. 1986b. Is there evidence for a 17 keV neutrino in the ^{35}S β spectrum? The case of Ohi et al. *Physics Letters* 174B:113–14.

Simpson, J. J., and A. Hime. 1989. Evidence of the 17-keV neutrino in the β spectrum of ^{35}S. *Physical Review D* 39:1825–36.

Smith, G. E. 1997. J.J. Thomson and the electron: 1897–1899 An introduction. *The Chemical Educator* 2:6.

Sommerfeld, A. 1916a. Zur Quantentheorie der Spectrallinien. *Annalen der Physik* 51:1–94.

——. 1916b. Zur Theorie des Zeeman-Effekys der Wasserstofflinien, mit einem Anhang über den Stark-Effeckt. *Physikalische Zeitschrift* 27:491–507.

Speake, C. C., T. M. Niebauer, M. P. McHugh, et al. 1990. Test of the inverse-square law of gravitation using the 300-m tower at Erie, Colorado. *Physical Review Letters* 65:1967–71.

Stacey, F. D., and G. J. Tuck. 1981. Geophysical evidence for non-Newtonian gravity. *Nature* 292:230–32.

Stacey, F. D., G. J. Tuck, S. C. Holding, et al. 1981. Constraint on the planetary scale value of the Newtonian gravitational constant from the gravity profile within a mine. *Physical Review D* 23:1683–92.

Staley, K. 1999. Golden events and statistics: What's wrong with Galison's image/logic distinction? *Perspectives on Science* 7:196–230.

Stoney, G. J. 1881. On the physical units of nature. *Philosophical Magazine* 11:381–390.

Stryer, L. 1975. *Biochemistry*. New York: W. H. Freeman.

Stubbs, C. W. 1990. Seeking new interactions: An assessment and overview. In *New and exotic phenomena '90: Tenth Moriond Workshop*, ed. O. Fackler and J. Tran Thanh Van, 175–85. Gif sur Yvette, France: Editions Frontieres.

Stubbs, C. W., E. G. Adelberger, B. R. Heckel, et al. 1989. Limits on composition-dependent interactions using a laboratory source: Is there a "fifth force?" *Physical Review Letters* 62:609–12.

Sur, B., E. B. Norman, K. T. Lesko, et al. 1991. Evidence for the emission of a 17-keV neutrino in the β decay of ^{14}C. *Physical Review Letters* 66:2444–47.

Thieberger, P. 1987. Search for a substance-dependent force with a new differential accelerometer. *Physical Review Letters* 58:1066–69.

———. 1989. Thieberger replies. *Physical Review Letters* 62:810.

Thomas, J., P. Kasameyer, O. Fackler, et al. 1989. Testing the inverse-square law of gravity on a 465m tower. *Physical Review Letters* 63:1902–5.

Thomson, G. P. 1928. The waves of an electron. *Nature* 122:279–82.

Thomson, J. J. 1897. Cathode rays. *Philosophical Magazine* 44:293–316.

Townsend, A. A. 1941. β-Ray spectra of light elements. *Proceedings of the Royal Society (London)* A177:357–66.

Tuck, G. J. 1989. Gravity gradients at Mount Isa and Hilton mines. Abstract from the Twelfth International Conference on General Relativity and Gravitation, Boulder, CO.

Tyler, A. W. 1939. The beta- and gamma- radiations from copper^{64} and europium^{152}. *Physical Review* 56:125–30.

Uhlenbeck, G. E., and S. Goudsmit. 1925. Ersetzung der Hypothese von unmechanischen Zwang durch eine Forderung bezuglich des inneren Verhaltens jedes einzelnen Elektrons. *Naturwissenschaften* 13:953–54.

———. 1926. Spinning electrons and the structure of spectra. *Nature* 117:264–65

van Fraassen, B. C. 1980. *The scientific image.* Oxford: Clarendon Press.

Villard, P. 1900. Sur la réflexion et la réfraction des rayons cathodiques et des rayon déviables du radium. *Comptes Rendus des Séances de L'Académie des Sciences* 130:1010–12.

von Baeyer, O., and O. Hahn. 1910. Magnetische Linienspektren von β-Strahlen. *Physikalische Zeitschrift* 11:488–93.

von Baeyer, O., O. Hahn, and L. Meitner. 1911. Über die β-Strahlen des aktiven Niederschlags des Thoriums. *Physikalische Zeitschrift* 12:273–79.

Watson, J. D. 1965. *Molecular biology of the gene.* New York: W. A. Benjamin.

Watson, J. D., and F. H. C. Crick. 1953a. Genetical implications of the structure of deoxyribonucleic acid. *Nature* 171:964–67.

———. 1953b. A structure for deoxyribose nucleic acid. *Nature* 171:737.

Weber, J. 1969. Evidence for discovery of gravitational radiation. *Physical Review Letters* 22:1320–24.

Weber, J., M. Lee, D. J. Gretz, et al. 1973. New gravitational radiation experiments. *Physical Review Letters* 31:779–83.

Weisberg, J. M., and J. L. Taylor. 1984. Observations of post-Newtonian timing effects in the binary pulsar PSR 1913 + 16. *Physical Review Letters* 52:1348–50.

Wigner, E. 1927. Einige Folgerungen aus der Schrödingerschen Theorie für die Termstrukturen. *Zeitschrift für Physik* 43:624–52.

Wilson, W. 1909. On the absorption of homogeneous β rays by matter, and on the variation of the absorption of the rays with velocity. *Proceedings of the Royal Society (London)* A82:612–28.

Wolfenstein, L. 1978. Neutrino oscillations in matter. *Physical Review D* 17:2369–74.

Wu, C. S., E. Ambler, R. W. Hayward, et al. 1957. Experimental test of parity nonconservation in beta decay. *Physical Review* 105:1413–15.

Index

fine structure constant, 121
Fischbach, Ephraim, 58, 59, 61, 63, 65, 66, 67, 72, 229
float experiment, 68–69, 71, 73
force: electromagnetic, 57, 170, 196, 197; fifth, 10, 57–78, 199, 228, 229; gravitational, 57, 198; magnetic, 16; nuclear 57; weak, 57
Friedman, Jerome, 21

G. See universal gravitational constant (G)
Galileo, 5, 62, 64
GALLEX experiment, 155–57, 162; II, 157
gallium (Ga): aqueous GaCl solution, 157; detector, 153–57
γ: cadmium capture, 122; and inverse β-decay, 120; particles, 99; radiation, 102, 107, 111; -rays, 165; scattering, 112
Garwin, Richard (Q), 199, 206, 208
Geiger counter, 23, 24, 107
germanium (Ge), 154, 155
GeV. See billion electron volts (GeV)
Giesel, Friedrich, 103
Gribov, V., 161
Glashow, Sheldon, 67
gold, 23
gravimeter, 74
gravitational field, 87
gravity: inverse squares, 67, 72; law of, 10, 57, 60, 61, 78; waves (radiation), 10, 193–209
Grodzins, Lee, 25, 26

hadrons, 212
Heisenberg uncertainty principle, 131
helium (He), 147, 149, 150; liquid, 173
Hertz, Heinrich, 83, 84
Hetherington, D., 138
high-energy physics, 6–7, 211, 215
Homestake mine experiment, 147–52, 162, 228
hybrid models, 123
hydrogen (H), 91; atom, 87, 92; burning of, 147; ion, 86, 87

icecap uniform density, 74, 75–76, 77
indium(^{114}In), 53, 54
infrared spectra, 4–5
interaction: electromagnetic, 16, 146, 171, 212; gravitational, 16; nuclear, 16, 146; weak (see neutrino, interaction with matter)

interferometer, 208
intermediate-range composition-dependent force, 63
intrinsic angular momentum, 98. See also electron; spin
intrinsic parity, 16–17
ion, 86, 87, 89
ionization, 88, 102, 106, 107, 120
irradiation, 147

Kamiokonde: super Kamiokonde detector, 165, 167; II water-Cerenkov counter experiment, 154, 156, 157–59, 161, 168 (see also Cerenkov, imaging water detector); III, 164
Kaufmann, Walter, 103–4
Kellogg, J. M. B., 122
Kelvin, Lord (William Thomson), 87
Kepler, Johannes: third law, 5
Kettlewell, Bernard, 7–9
keV. See 17-keV
kink in decay energy spectrum, 133, 134, 135, 137, 140, 141, 145
Konopinski, Emil, 10, 44, 46, 55, 56. See also Konopinski-Uhlenbeck theory of β-decay
Konopinski-Uhlenbeck theory of β-decay, 10, 46–56, 229
Kurie, F., 48, 50; plot, 46, 48, 50, 51, 53, 133
Kuz'min, V. A., 152

lasers, 194–95, 214
lead, 98, 228; glass shower counter, 221; ^{214}Pb (radium B), 105, 107
Lee, Tsung Dao, 15–16, 17, 18, 21, 31
left-right symmetry. See mirror-reflection symmetry
Lenard, Philipp, 86, 105
lepton, 128, 216. See also τ-lepton
light: leptons, 128 (see also τ-lepton); and motion of earth, 169; and organic liquids, 119–20; particle and wave characteristics, 22; polarized, 212; quantum of energy, 91; speed of, 171
linear law, 106
local mass asymmetry, 65, 66, 67, 68
loop detector, 174, 175, 176
Lorentz, Hendrick, 103, 104; Lorentz-Fitzgerald contraction, 169–70

macromolecules, 34
magnetic deflection, 83, 84, 85–86

magnetic field, 93, 107, 171
magnetic flux, 172, 174, 175, 176
magnetometer. See SQUID (superconduc-
tion quantum interference device)
magnetometer
magnetons, 93, 95
mass: gravitational, 62, 63; inertial, 61–62
mass-to-charge ratio (m/e). See under
cathode rays
Meselson, Matthew, 34, 35, 41, 43
Meselson-Stahl experiment, 34–43
meson, 126; decay, 18, 58
MeV. See million electron volts
Meyer, Stefan, 103
Michelson-Morley experiment, 169–70
Millikan, Robert, 10, 87, 88–90, 91–92; and
the electron charge, 183–92, 227–28
million electron volts (MeV), 51, 52, 122,
147, 148, 157; MeV/c, 126. See also 17-
keV; billion electron volts (GeV)
mineshaft gravity, 61, 63, 65, 66, 74, 76, 77
mirror-reflection symmetry, 10, 15–16, 17,
31
monopoles, magnetic, 10, 170–79, 229
moth. See peppered moth
motion detector, 176, 177, 179
Mott, Nevill F., 22, 23, 26, 29, 30, 227
MSW (Mikheyev, Smirnov, and Wolfenstein) effect, 161
μ/τ events, 164–65
muon (μ), 18–19; cosmic ray, 164, 165;
decay, 160, 165 (see also under neutrino); polarization, 162, 163; tri-, 216

nanoamperes (nA), 175
neutral current (NC), 167, 168; interactions, 211, 212, 213, 215, 216, 217
neutrino, 18, 45, 52, 98–145, 147, 215,
228–29; antineutrino, 119, 122, 124, 160,
215; atmospheric, 165; created from nucleus disintegration, 117, 122; detection
of, 120–23, 146; detector, 150; discovery
of by Chadwick, 116; electron, 126, 128,
130, 166, 167; handedness, 99, 229;
heavy, 133, 134, 137, 140; high-energy,
125–26, 130; interaction with matter,
98, 116, 127, 147, 215; kinds of, 99,
124–31; mass, 99, 132, 228; muon, 124,
125–28, 130, 160, 162, 163, 165; as neutral,
99; oscillations, 159, 160–66, 167, 228,
229; and pion decay, 125–26, 128; and
the positron (e⁺), 119, 120, 122; p-p,

155–56; and protons, 162; solar, 10,
146–68, 228; and solar flux, 149, 160,
167, 168; tau (see τ-lepton); thousand
electron volt (17-keV), 132–45, 228, 229;
transformation, 99, 228
neutron, 45, 63, 115, 116, 122, 126; beam and
gravity, 58. See also τ-lepton; under
electron
Newton, Isaac, 10, 74
nickel (⁶³Ni) spectrum measurement, 140
nitrogen, liquid, 173, 177
nitrogen isotope (¹⁴N and ¹⁵N), 34, 35, 37,
39, 40, 42
non-radiating electronic orbits, 91, 92
Novosibirsk experiment, 216, 217, 218
nuclear fission, 119, 122, 124

oil drop experiment, 88–90. See also Millikan, Robert, and the electron charge
1/α. See fine structure constant
optical rotation, 213
orbital angular momentum, 92, 93
oriented nuclei, 18, 19, 21

parity conservation: in quantum mechanics, 16; violation of, 15–31; and weak
reactions, 15, 16, 17–18, 20, 21–22, 25
parity nonconservation, 10, 15, 19–20, 23,
215, 228. See also parity conservation,
violation of
particles, 22, 58, 63, 87, 99, 229; as constituents of atoms, 86, 91; cosmic, 123;
elementary (see Z⁰ boson); light (see
leptons); neutral, 99, 112
Pauli, Wolfgang, 16, 17–18, 21, 45, 94, 115–16,
121, 124, 228
peppered moth, 7–9
perchlorethylene/tetrachloroethylene
(C₂Cl₄), 148, 149–50
Perrin, Jean, 82, 83
phosphorescence, 83, 100–101
phosphorus, 46, 51, 52, 53, 55
photoelectric effect, 22
photomultiplier tube, 120, 162
photon: 146; showers, 126
π mesons. See pion
Pickering, Andrew, and atomic parity
violation, 210–12, 213–14, 215, 216, 217,
218–20, 223, 225, 226
piezo-electric crystals, 196, 198
pion, 58, 124, 125–26, 162; neutral, 165
Planck's constant, 91, 171, 172

Stoney, G. Johnstone, 87
strain gauge, 176, 177, 198
submarine gravity, 61
Sudbury Neutrino Observatory (SNO),
162, 166-67, 168, 229
sulfur (^{35}S) β-decay energy, 136, 138, 140
superconducting lead shield, 176
supernova (SN 1987A), 160
symmetry. *See* mirror-reflection symmetry

τ-lepton, 129-31, 165
TEA (transverse excited atmosphere) laser,
194, 195
Telegdi, Valentine, 21
telescope, 5, 6
tetrachloroethylene liquid (C_2Cl_4), 149-50
Tevatron, 130
thallium, 217
theory. *See under* experiment, and theory
θ, 19
Thieberger, Peter, 69, 71, 73
third particle, 45, 132
Thomsen, Joseph John, 81-82, 83, 84-85, 87
thousand electron volts. *See* 17-keV
torsion pendulum, 62, 66, 68, 70, 71, 72
tower gravity, 74, 75, 76, 77
tritium, 133-34; decay energy, 134, 135, 136,
138

Uhlenbeck, George, 46, 56. *See also*
Konopinski-Uhlenbeck theory of
β-decay
uncertainty, 131, 192, 225-26
unified theory of electroweak interactions.
See Weinberg-Salam (W-S) unified
theory of electroweak interactions

universal gravitational constant (G), 60-61,
62, 63, 74
τ-θ puzzle, 16-17, 28
uranium: nitrate, 101; radioactivity of, 102;
salts, 100, 101-2

validation. *See under* experiment
vertex, 165
Villard, Paul, 102

Washington-Oxford experiments, 211,
212-16, 218, 225
water: mass of, 72; ultrapure heavy (D_2O),
167
Watson, James, 32, 33, 39, 40, 41-42
wave: equations, 16; function, 17
weak reactions. *See under* parity conservation
Weber, Joseph, 195-98, 199, 200-203, 205-7,
209; gravity wave detector, 195-96
Weinberg-Salam (W-S) unified theory of
electroweak interactions, 6, 10, 211, 212,
215, 216, 217, 218, 223, 226
Weyl, Herman, 16
Wigner, Eugene, 16
Wilkins, Maurice, 42
Wilson experiment, 106-7
Wooster, William, 112-13, 114
Wu, Chien-Shiung, 19, 21

x-radiation, 88
x-rays, 22, 100

Yang, Chen Ning, 15-16, 17, 19, 21, 31

Z^0 boson (elementary particle), 130-31